Glencoe Science

LAB MANUAL

Biology

Biology online
glencoe.com

Mc Graw Hill **Glencoe**

New York, New York   Columbus, Ohio   Chicago, Illinois   Peoria, Illinois   Woodland Hills, California

Glencoe

The McGraw·Hill Companies

Send all inquiries to:
Glencoe/McGraw-Hill
8787 Orion Place
Columbus, OH 43240-4027

ISBN-13: 978-0-07-874720-5
ISBN-10: 0-07-874720-1

Printed in the United States of America.

2 3 4 5 6 7 8 9 10 11 045 11 10 09 08 07 06

Table of Contents

Table of Contents, continued

How to Use This Laboratory Manual

Working in the laboratory throughout the course of the year can be an enjoyable part of your biology experience. This laboratory manual is a tool for making your laboratory work both worthwhile and fun. The laboratory activities are designed to fulfill the following purposes:

- to stimulate your interest in science in general and especially in biology
- to reinforce important concepts studied in your textbook
- to allow you to verify some of the scientific information learned during your biology course
- to allow you to discover for yourself biological concepts and ideas not necessarily covered in class or in the textbook readings
- to acquaint you with a variety of modern tools and techniques used by today's biological scientists

Most importantly, the laboratory activities will give you firsthand experience in how a scientist works.

The activities in this laboratory manual are of two-types: Classic or Design Your Own. In a *Classic* activity, you will be presented with a problem and will use the steps of the experiments to draw conclusions. In *Design Your Own* activities, you will be given background information, and then will be asked to develop your own hypothesis and design activities and evaluation procedures to test it. In both kinds of activities, you will need to use scientific methods to obtain data and answer questions.

The basic format for the activities is described below. Understanding the purpose of each section will help guide you as you work through each activity.

Introduction: A brief introduction provides background information for each activity. You might need to refer to the introduction for information that is important for completing an activity.

Objectives: The list of objectives is a guide to what will be done in the activity and what will be expected of you.

Materials: The materials section lists the supplies you will need to complete the activity. Check with your teacher to obtain these materials.

Procedure: (*Classic* activities) The procedure gives you step-by-step instructions for carrying out the activity. Many steps have safety precautions. Be sure to read these statements and obey them for your own and your classmates' protection. Unless told to do otherwise, you are expected to complete all parts of each assigned activity. Important information needed for the procedure, but that is not an actual procedural step, is also found in this section.

Hypothesis: (*Design Your Own* activities) You will write a hypothesis statement to express your expectations of the results and as a response to the problem statement.

Plan the Experiment: (*Design Your Own* activities) In this section, you will plan how to obtain data, guided by the background information provided to you.

Check the Plan: (*Design Your Own* activities) Have your procedure approved by the teacher before proceeding.

Record the Plan: (*Design Your Own* activities) Write your experimental plan, and sketch your equipment setup.

Data and Observations: This section includes tables and space to record data and observations.

Analyze and Conclude: In this section, you will draw conclusions about the results of the activity just completed. Rereading the introduction before answering the questions might be helpful.

Write and Discuss: (*Design Your Own* activities) This section provides material you might use in a classroom discussion or homework assignment based on the activity.

Inquiry Extensions: This section includes ideas for ways to extend the activity or plan related experiments.

In addition to the activities, this laboratory manual has several other features—a description of how to write a lab report, a section on the care of living things, diagrams of laboratory equipment, and information on safety that includes first aid and a safety contract. Read the section on safety now. Safety in the laboratory is your responsibility. Working in the laboratory can be a safe and fun learning experience and can help you to understand and enjoy biology.

Writing a Laboratory Report

When scientists perform experiments, they make observations, collect and analyze data, and formulate generalizations about the data. When you work in the laboratory, you should record all your data in a laboratory report. An analysis of data is easier if all data are recorded in an organized, logical manner. Tables and graphs are often used for this purpose. A written laboratory report should include all of the following elements.

TITLE: The title should clearly describe the topic of the report.

HYPOTHESIS: Write a statement to express your expectations of the results and as an answer to the problem statement.

MATERIALS: List all laboratory equipment and other materials needed to perform the experiment.

PROCEDURE: Describe each step of the procedure so that someone else could perform the experiment following your directions.

RESULTS: Include in your report all data, tables, graphs, and sketches used to arrive at your conclusions.

CONCLUSIONS: Record your conclusions in a paragraph at the end of your report. Your conclusions should be an analysis of your collected data.

Read the following description of an experiment, then answer the questions.

All plants need water, minerals, carbon dioxide, sunlight, and living space. If these needs are not met, plants cannot grow properly. A biologist thought that plants would not grow well if too many were planted in a limited area. To test this idea, the biologist set up an experiment. Three containers were filled with equal amounts of potting soil. One bean seed was planted in Container 1, five seeds in Container 2, and ten seeds in Container 3. All three containers were placed in a well-lit room. Each container received the same amount of water every day for two weeks. The biologist measured the heights of the growing plants every day. Then the average height of the plants in each container each day was calculated and recorded in a table. The biologist then plotted the data on a graph.

1. What was the purpose of this experiment?

_____

2. What materials were needed for this experiment?

_____

3. Write a step-by-step procedure for this experiment.

_____

_____

_____

_____

_____

_____

_____

_____

4. Table 1 shows the data collected in this experiment. Based on these data, state a conclusion for this experiment.

_____

_____

_____

_____

Table 1

Average Height of Growing Plants (mm)										
	Day									
Container	1	2	3	4	5	6	7	8	9	10
1	20	50	58	60	75	80	85	90	110	120
2	16	30	41	50	58	70	75	80	100	108
3	10	12	20	24	30	35	42	50	58	60

5. Plot the data in **Table 1** on a graph. Show average height on the vertical axis and the days on the horizontal axis. Use a different colored pencil to graph the results of each container.

Care of Living Things

Caring for living things in a biology laboratory can be interesting and fun, and it can help develop the respect for all life that comes only from firsthand experience. In a room with an aquarium, terrarium, healthy animals, or growing plants, there is always some observable interaction between organisms and their environment. There are many species of plants and animals that are suitable for a classroom, but having them should be considered only if proper care will be taken so that the organisms not only survive, but thrive. Before growing plants or bringing animals into a classroom, find out if there are any health or safety regulations restricting their use, or if there are any applicable state or local laws governing live plants and animals. Also, do not consider cultivating any endangered or poisonous species. A biological supply house or local pet store will provide growing tips for plants or literature on animal care when these organisms are purchased.

Evaluating Resources

Before bringing any live specimens into a new environment, check with your teacher to see if their basic needs will be met in their new location. Plants need either sunlight or grow lights. Animals must be placed in well-ventilated areas out of direct sunlight and away from the draft of open windows, radiators, and air conditioners. For both animals and plants, a source of fresh water is essential. Consider the likely fluctuation in temperature over weekends and holidays, and who will care for the plants or animals during those times.

Setting Up an Aquarium

A closed system such as an aquarium supports a variety of animals and plants and can be maintained easily if set up correctly. A 10- or 20-gallon tank can be a suitable home for about 5 to 10 tropical fish or even more of the temperate goldfish. An air pump, filter, heater, thermometer, and aquarium light (optional) need to be in working order.

First fill an aquarium with a layer of gravel, then fill with water. If using tap water, let the water stand a day before putting any fish in the tank. During cooler months, adjust the thermostat of the heater to bring the water to the desired temperature before adding fish. Most fish require temperatures of 20° to 25°C. An inexpensive pH kit purchased from a pet store will test the acidity of the water and guide the maintenance of a healthy pH.

Choose fish that are compatible with one another. A pet-store clerk can help in the selection. It is worth purchasing a scavenger fish, such as a catfish, or an algae eater that will help keep the tank clean of algae. Snails are also helpful for this purpose. After purchasing, keep fish in the plastic bag containing water in which they came. Float the bag in the aquarium until the water reaches the same temperature, then slowly let the fish swim out of the bag. Some fish, such as guppies, eat their young. A smaller brood tank can be placed inside the aquarium to keep the mother separated from the young.

One person should be responsible for feeding the fish. Feed fish sparingly. Overfeeding is not healthy for the fish; also, it clouds the tank and causes unnecessary decay. Weekend or vacation food should also be available. These are slow-dissolving tablets that can feed the fish over vacations.

Plants can be added to an aquarium as well. *Elodea, Anacharia, Sagittaria, Cabomba,* or *Vallianeria* grown in a fish tank also are useful for many biology lab activities. Monitor their growth carefully and trim plants if growth is excessive. Some fish and snails might nibble on the plants, causing them to break apart and decay. Decay introduces bacterial populations that can endanger the fish, so be sure to remove any decaying plant matter.

Variations on an aquarium include setting up a "balanced aquarium" with fish, plants, and scavengers in balance so that no pump or filter is necessary. This usually takes more planning and maintenance than a filtered tank. More maintenance is needed also for a marine aquarium because of the corrosive nature of salt water. However, if specimens of marine organisms are readily available, creating such a mini-habitat is well worth the effort.

Keeping Mammals in a Classroom

Keeping mammals takes more consideration and commitment of time and expense. A small mammal such as a gerbil, guinea pig, hamster, or rabbit can be kept in a classroom, at least for a short time. Explore the possibility of dwarf breeds that are more at home in a small space. However, many mammals are sensitive, social mammals that form bonds and attachments to people. Life in a small cage alone most of the time is not suitable for a long and healthy life. For short periods of time, however, small animals can be kept in a cage, provided it is clean and large enough. Find out the exact nutritional needs of the animals; feed them on a regular schedule and provide fresh water daily. Some animals require dry food supplemented with fresh foods, such as greens. However, these foods spoil more rapidly, and uneaten portions must be removed. Provide a cage large enough for the animal, as well as materials for bedding, nesting, and gnawing. Clean the cages frequently. Letting urine and feces collect in a cage fosters the growth of harmful bacteria. Animals in a cage also require an exercise wheel. Lack of space combined with overeating can make an animal overweight and lethargic. Handle animals gently. Under no circumstances should animals be exposed to harmful radiation, drugs, toxic chemicals, or surgical procedures.

Many times students want to bring a pet or even a wild animal that they have found into the classroom for observation. Do so only with discretion and if a proper cage is available. Protective gloves and glasses should be worn while handling any animals with the potential to bite. Be sure to check with local park rangers or wildlife specialists for any wildlife restrictions that may apply. Return any wild animals to their environment as soon as possible after observations.

Growing Plants in the Classroom

To successfully grow plants in a classroom, have on hand commercial potting soil, suitable containers such as clay or plastic pots, plant fertilizer, a watering can, and a spray bottle for misting. Always put a plant in the correct size container. One that is too large will encourage root growth at the expense of the stem and leaves. Place bits of broken clay or gravel in the bottom of the pot for drainage, then add potting soil and the plant. Place in a warm, well-lighted area and supply water. Give careful attention to a new plant to assess its adaptation to its new environment. Pale leaves might indicate insufficient light, yellowing

leaves indicate overwatering, and dropping leaves usually indicate insufficient humidity. Fertilize only as directed.

With little special attention, plants such as geraniums, begonias, and coleus can be easily and inexpensively grown in a classroom. These plants are hardy and can withstand fluctuations in light and temperature. From one hardy plant, many cuttings can be made to demonstrate vegetative propagation. A cutting of only a few leaves on a stem will develop roots in 2 to 4 weeks if it is placed in water or given root-growth hormone powder.

These plants not only add color to a classroom but are useful in biology experiments as well. The dense green leaves of geraniums are especially useful for extracting chlorophyll or showing the effects of light deprivation. The white portions of variegated coleus leaves are good for showing the absence of photosynthesis with a negative starch-iodine reaction. Pinch back the flower buds as they begin to form to encourage fuller leaf growth.

Larger plants such as a fig (*Ficus*), dumbcane (*Dieffenbachia*), cornplant (*Dracaena*), Norfolk Island pine (*Araucaria*), umbrella plant (*Schefflera*), or various philodendrons adapt well to low-light conditions and so do not need frequent watering. However, make sure humidity is suitable to avoid dropping leaves. More exotic plants might be best suited to a small-dish garden but will need special care because there is less soil to hold moisture.

During winter months, a dish garden of forced bulbs, such as paperwhite narcissus, can be easily grown by placing the bulbs in a container of water left in a cool, dark place. Blooms will appear in 3 to 4 weeks. In the early spring, shoots of early flowering shrubs, such as forsythia and pussy willow, may be forced. Cut off some healthy shoots when buds appear, wrap in wet newspaper, then bring indoors and immerse cut ends in a tall vase or jar. Also buds of fruit trees, such as apple, plum, or peach, will produce leaves and flowers in this way. Be sure to maintain shoots by changing water when necessary.

Laboratory Equipment

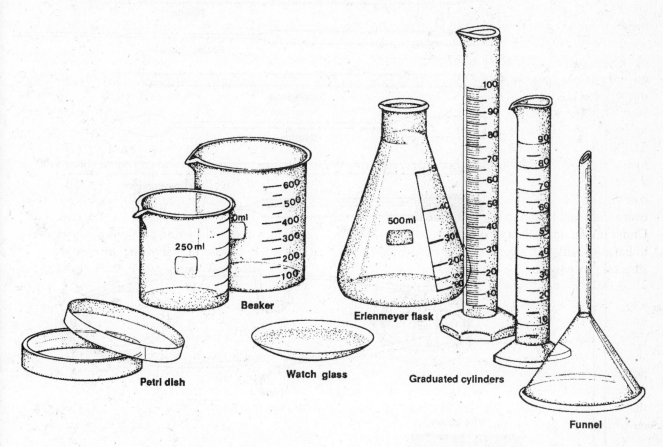

Beaker

250 ml

Erlenmeyer flask

500ml

Graduated cylinders

Funnel

Petri dish

Watch glass

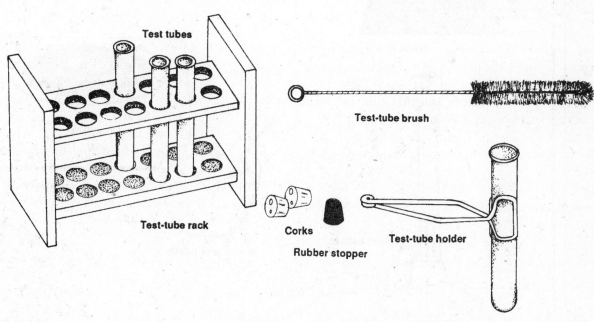

Test tubes

Test-tube brush

Test-tube rack

Corks

Rubber stopper

Test-tube holder

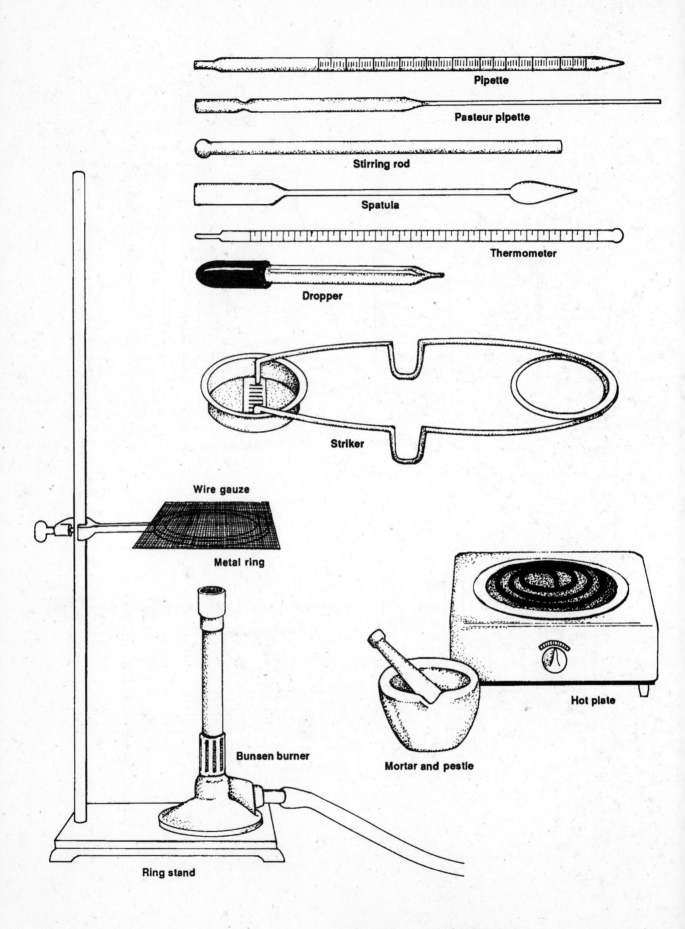

Pipette

Pasteur pipette

Stirring rod

Spatula

Thermometer

Dropper

Striker

Wire gauze

Metal ring

Hot plate

Bunsen burner

Mortar and pestle

Ring stand

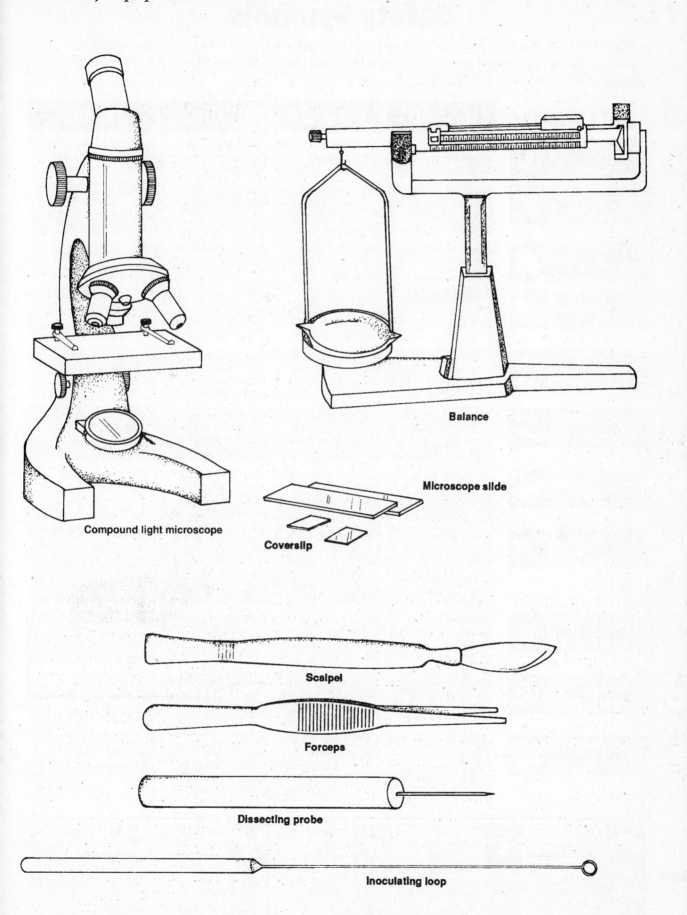

Compound light microscope

Balance

Microscope slide

Coverslip

Scalpel

Forceps

Dissecting probe

Inoculating loop

Safety Symbols

These safety symbols are used in laboratory and field investigations in this book to indicate possible hazards. Learn the meaning of each symbol and refer to this page often. *Remember to wash your hands thoroughly after completing lab procedures.*

SAFETY SYMBOLS	HAZARD	EXAMPLES	PRECAUTION	REMEDY
DISPOSAL	Special disposal procedures need to be followed.	certain chemicals, living organisms	Do not dispose of these materials in the sink or trash can.	Dispose of wastes as directed by your teacher.
BIOLOGICAL	Organisms or other biological materials that might be harmful to humans	bacteria, fungi, blood, unpreserved tissues, plant materials	Avoid skin contact with these materials. Wear mask or gloves.	Notify your teacher if you suspect contact with material. Wash hands thoroughly.
EXTREME TEMPERATURE	Objects that can burn skin by being too cold or too hot	boiling liquids, hot plates, dry ice, liquid nitrogen	Use proper protection when handling.	Go to your teacher for first aid.
SHARP OBJECT	Use of tools or glassware that can easily puncture or slice skin	razor blades, pins, scalpels, pointed tools, dissecting probes, broken glass	Practice common-sense behavior and follow guidelines for use of the tool.	Go to your teacher for first aid.
FUME	Possible danger to respiratory tract from fumes	ammonia, acetone, nail polish remover, heated sulfur, moth balls	Make sure there is good ventilation. Never smell fumes directly. Wear a mask.	Leave foul area and notify your teacher immediately.
ELECTRICAL	Possible danger from electrical shock or burn	improper grounding, liquid spills, short circuits, exposed wires	Double-check setup with teacher. Check condition of wires and apparatus.	Do not attempt to fix electrical problems. Notify your teacher immediately.
IRRITANT	Substances that can irritate the skin or mucous membranes of the respiratory tract	pollen, moth balls, steel wool, fiberglass, potassium permanganate	Wear dust mask and gloves. Practice extra care when handling these materials.	Go to your teacher for first aid.
CHEMICAL	Chemicals that can react with and destroy tissue and other materials	bleaches such as hydrogen peroxide; acids such as sulfuric acid, hydrochloric acid; bases such as ammonia, sodium hydroxide	Wear goggles, gloves, and an apron.	Immediately flush the affected area with water and notify your teacher.
TOXIC	Substance may be poisonous if touched, inhaled, or swallowed.	mercury, many metal compounds, iodine, poinsettia plant parts	Follow your teacher's instructions.	Always wash hands thoroughly after use. Go to your teacher for first aid.
FLAMMABLE	Open flame may ignite flammable chemicals, loose clothing, or hair.	alcohol, kerosene, potassium permanganate, hair, clothing	Avoid open flames and heat when using flammable chemicals.	Notify your teacher immediately. Use fire safety equipment if applicable.
OPEN FLAME	Open flame in use, may cause fire.	hair, clothing, paper, synthetic materials	Tie back hair and loose clothing. Follow teacher's instructions on lighting and extinguishing flames.	Always wash hands thoroughly after use. Go to your teacher for first aid.

 Eye Safety Proper eye protection should be worn at all times by anyone performing or observing science activities.

 Clothing Protection This symbol appears when substances could stain or burn clothing.

 Animal Safety This symbol appears when safety of animals and students must be ensured.

 Radioactivity This symbol appears when radioactive materials are used.

 Handwashing After the lab, wash hands with soap and water before removing goggles

Student Lab/Activity Safety Form

Student Name: _____

Date: _____

Lab/Activity Title: _____

In order to show your teacher that you understand the safety concerns of this lab/activity, the following questions must be answered after the teacher explains the information to you. You must have your teacher initial this form before you can proceed with the activity/lab.

1. How would you describe what you will be doing during this lab/activity?

2. What are the safety concerns associated with this lab/activity (as explained by your teacher)?

- _____
- _____
- _____
- _____
- _____

3. What additional safety concerns or questions do you have?

Adapted from Gerlovich, et al. (2004). The Total Science Safety System CD, JaKel, Inc.
Used with Permission.

Design Your Own
Lab 1

What makes mold grow?

Have you ever opened a bag of bread and found green or white mold growing on the bread? Where did this mold come from? What types of conditions are better for mold growth, and what measures can you take to avoid them? In this laboratory exercise, you will design an experiment to test one of the conditions that might result in bread growing mold.

Problem
Determine what conditions are ripe for the growth of mold on bread.

Objectives
- Write a hypothesis.
- Develop an experiment to test the hypothesis.
- Control variables during the experiment.
- Draw conclusions about the formation of mold on bread.

Safety Precautions

WARNING: *Do not eat any food in a science lab. Do not open the sealed bags. The release of mold spores can aggravate allergies, asthma, and other medical conditions.*

Possible Materials
paper plates
dropper
bread (with no preservatives)
plastic bags (sealable)
tap water
tape

Hypothesis
Use what you know about the mold found on bread to write a hypothesis indicating what factors influence the formation of mold.

_____

_____

_____

_____

Design Your Own **Lab** 1, **What makes mold grow?** continued

Plan the Experiment

1. Read and complete the lab safety form.
2. Make a list of the factors that might influence the formation of mold on bread. Be sure to test the factors you listed in your hypothesis.
3. Decide on a procedure for testing your hypothesis. In the space provided, write your procedure for testing the factors. Include a list of the materials you will use.
4. Identify the independent variable, dependent variable, constants, and control group.
5. Decide how you will record your data and when you will record it. Design a data table to collect information about the presence of mold over a period of six days. Be sure to collect quantitative data that can answer these types of questions: How many colonies are there? What is the size of each colony?

Check the Plan

1. Be sure that a control group is included in your experiment and that the experimental groups vary in only one way.
2. Make sure your teacher has approved your experimental plan before you proceed.
3. When you have completed your experiment, dispose of materials as directed by your teacher.

Record the Plan

In the space below, write your experimental procedure and make a sketch of your experimental setup.

Design Your Own **Lab** 1, **What makes mold grow?** continued

Data and Observations

1. Use the space below to create a data table of your findings, including information about the presence of mold.

Analyze and Conclude

1. How did the appearance of the two slices of bread change over the six days?

_____

_____

_____

2. How can you explain differences in the appearance of the bread?

_____

_____

_____

3. What was the manipulated variable in your experiment? Why was it necessary to control all the other variables other than this one?

_____

_____

_____

4. Describe the control in your experiment. What did the control show?

_____

_____

_____

_____

5. Error Analysis What were some possible sources of error in your experiment?

_____

_____

_____

6. Exchange your procedure and data with another group in your class for peer review. Discuss any differences in the results.

_____

_____

_____

Write and Discuss

Write a short paragraph describing your findings and indicating whether or not they support your hypothesis. Discuss any questions your findings might have raised.

_____

_____

_____

_____

Inquiry Extensions

1. Many health-food stores and supermarkets now sell organic baked goods with no preservatives, but many brands of bread continue to use them. How well do these preservatives work at reducing the time it takes for mold to form? Design an experiment that tests the differences between the formation of mold on bread with preservatives and organic bread.

2. What other conditions could influence the rate of mold formation? Temperature? Exposure to sunlight v. artificial light? Contact with other foods? Design an experiment to test one of these hypotheses, or a hypothesis that you develop on your own, and report your results to the class.

Design Your Own
Lab 2

How does your biome grow?

The environmental factors that affect the growth of an organism can be grouped into two categories—biotic and abiotic. Biotic factors are living organisms in the environment. Abiotic factors include naturally occurring substances in the soil, such as chemicals and nutrients, as well as water, sunlight, and temperature. In this lab, you will create a model biome and study the effects of abiotic factors on germinating plants.

Problem
What impact do abiotic factors have on biomes?

Objectives
- Form a hypothesis about the impact of abiotic factors on a biome.
- Design an experiment to test your hypothesis.
- Identify a control to the experiment.
- Make a model of a biome.
- Create a data table.
- Draw conclusions.

Safety Precautions
Wash your hands thoroughly with soap and water after handling the soil.

Possible Materials
bicarbonate of soda tablets
clear plastic bottles (2-L soda bottles)
clear plastic wrap
colored gels or mylar
electric fan
flower seeds
grass seeds
lima bean seeds
index cards
lamps
masking tape
sterile potting soil
alternative soil types (sand, clay, loam)
scissors
small rocks
small beaker or test tubes
tape
water

Hypothesis
Use what you know about ecosystems and ecology to write a hypothesis indicating the effect of an abiotic factor of your choice on the germination of plants in a model biome.

_____

_____

_____

_____

Design Your Own **Lab** 2, **How does your biome grow?** continued

Plan the Experiment

1. Read and complete the lab safety form.
2. Choose which biome you wish to simulate. Be sure that your biome is indicated in your hypothesis.
3. Decide on a procedure to use to test the impact of an abiotic factor on your simulated biome.
4. Identify the independent variable, dependent variable, constants, and control group.
5. Describe how you will measure and record your data.

Check the Plan

1. Make sure your teacher has approved your experimental plan before you proceed.
2. Be sure that a control group is included in your experiment and that the experimental group varies in only one way.
3. Observe and record the impact of abiotic factors on the biotic components of your simulated biome. Be sure to make sketches each day of your biome and the changes you observe. Be detailed in your drawings. Provide quantitative observations (using measurements).
4. When you have completed the experiment, ask your teacher whether you should continue to make long-term observations or dispose of the organisms as he or she directs.

Record the Plan

In the space below, write your experimental procedure and make a sketch of your experimental design.

Design Your Own **Lab** 2, **How does your biome grow?** continued

Data and Observations

1. Use the space below to create a data table of your findings.

Analyze and Conclude

1. On which abiotic factor did you focus? Why?

_____

_____

_____

2. Did this abiotic factor seem to have a significant impact on the dependent variable in your simulated ecosystem? Explain.

_____

_____

_____

3. Describe the control in your experiment. What was held constant in the control? Why was it set up that way?

_____

_____

_____

_____

Design Your Own **Lab** 2, **How does your biome grow?** continued

4. How does your experiment relate to biomes and abiotic factors in nature?

_____

_____

5. **Error Analysis** What are some possible sources of error in your experiment?

_____

_____

6. Exchange your procedure and data with another group in your class. What do their data show about the biome they chose to simulate? What conclusions can you draw about the abiotic factors in a biome?

_____

_____

7. What are the limitations of the design of this experiment? Are there additional factors at work?

_____

_____

Write and Discuss
Write a short paragraph describing your findings and indicating whether or not they support your hypothesis. Discuss any questions your results have raised.

_____

_____

_____

_____

_____

_____

Inquiry Extensions
1. Describe the rainfall pattern and abiotic factors that make up the biome you live in. How do these factors impact the plants, animals, and agriculture in your area?
2. If you were to maintain your biomes in the classroom or at home, what abiotic factors would you change from your original model? Make a prediction about what you would observe under the new conditions.

Design Your Own
Lab 3

Do freshwater biomes respond differently to acid rain?

Your state happens to be downwind from a volcano that recently ejected large amounts of ash and sulfur and, as a result, rainwater in your area has been unusually acidic. You are conducting research for your state's water resources department to see how different types of lakes are affected by this acidic rainwater.

Problem

Design an experiment that will test how acid rain impacts algae in a freshwater ecosystem. You might want to explore a variety of ecosystems found in different areas of the country or your state, such as a pond or lake with limestone bedrock or a pond or lake with granite bedrock.

Objectives

- Design an experiment to assess the impact of acid rain on a freshwater pond.
- Conduct the experiment and record data.
- Interpret data and draw conclusions.

Safety Precautions

WARNING: *Use caution with the acid rain samples—they might cause chemical burns. Be sure to keep lamps away from water sources to avoid a potential shock hazard.*

Possible Materials

pond water
algae samples
limestone chips
acid rain sample
large glass jars (2)
granite gravel
pH test strips
large dropper or pipette
light source

Hypothesis

Use what you know about pH and acid rain to write a hypothesis indicating what impact acid rain will have on an aquatic environment.

_____

_____

_____

_____

Plan the Experiment

1. Read and complete the lab safety form.
2. Choose the type of gravel you will use to line your pond: granite or limestone.
3. Decide on a procedure to use when setting up your pond environment and for testing the pH of the water.
4. Identify the independent variable, dependent variable, constants, and control group.
5. Decide how you will record your data and when you will record it. Design a data table to collect information about the pH of the water, the time that has elapsed, and the status of the algae growing and living in your pond.
6. Determine the length of time needed to observe your samples.

Check the Plan

1. Be sure that a control group is included in your experiment and that the experimental group varies in only one way.
2. Make sure the teacher has approved your experimental plan before you proceed.
3. Observe the impact that acid rain has on the growth of algae in your simulated pond environment.
4. When you have completed the experiment, dispose of the pond water as instructed by your teacher. Be sure to wash your hands with soap and water after you are done.

Record the Plan

In the space below, write your experimental procedure and make a sketch of your experimental setup.

Name _____ Date _____ Class _____

Design Your Own **Lab** 3, **Do freshwater biomes respond differently to acid rain?** continued

Data and Observations

1. Use the space below to create a data table of your findings, including the length of time that has passed, the pH of the water each day, and the status of the algae that is growing in the pond water.

Analyze and Conclude

1. How did the pH of your simulated pond change from day to day?

_____

_____

_____

_____

2. What impact did the stone lining at the bottom of your pond have on the pH of the water?

_____

_____

_____

_____

3. How did the algae in your pond survive? Describe any changes in the appearance of the algae. Explain possible reasons for what you found.

_____

_____

_____

_____

4. What was the control in your experiment? What did the control show?

_____

_____

_____

_____

5. Error Analysis What were some possible sources of error in your experiment?

_____

_____

_____

_____

6. For the peer review process, exchange your procedure and data with a group that used the same stone as you did and with a group that used the other option. Does comparing your data with the data from the other groups indicate that the presence of granite as opposed to limestone can affect the pH in a pond environment? What conclusion can you draw?

_____

_____

_____

_____

Write and Discuss
Write a short paragraph describing your findings and indicating whether or not they support your hypothesis. Discuss any questions your results might have raised.

_____

_____

_____

_____

_____

Inquiry Extensions
1. How does temperature impact the pH of a sample? Design an experiment that examines the impact of temperature on the pH of pond water.

2. Why should you be concerned about the impact of acid rain on ponds, lakes, and streams? How does the pH of pond water ultimately impact your life as a teenager in the United States?

Classic
Lab 4

How can you show a population trend?

Populations exhibit growth because of births as well as immigration. The rate at which populations grow depends on many factors. One factor is the rate at which the species can produce offspring. Factors that limit populations, called limiting factors, include predators, disease, food supply, and availability of suitable habitat.

Some of these factors depend on the density of the population, while others do not. Density-dependent factors include the availability of food, the occurrence of disease, stress, light availability, and predators. Density-independent factors include extreme weather, fires, seasonal changes, floods, and changes to habitat (such as tree cutting).

Objectives

- Culture bacterial colonies to track population growth.
- Graph population data, choosing appropriate scales and titles.
- Compare and contrast populations and the factors that affect growth.

Materials

pieces of graph paper (4)
ruler
pencil
eraser
calculator
colored pencils
petri dishes with lids (2)
nutrient agar
masking tape
permanent marker

Safety Precautions

WARNING: *Use protective gloves if the petri dishes were recently removed from the autoclave—they might be hot. Always wash your hands after handling the cultured petri dishes.*

Procedure

Part A. Counting a Bacterial Population

1. Read and complete the lab safety form.
2. Prepare two growth chambers. Sterilize two petri dishes and their tops.
3. Place the agar into both petri dishes. Cover one of the dishes, and seal it with tape. This will be your control. Using the marker, label a piece of masking tape with the letter *A*, and affix it to the bottom of the dish.
4. Run your fingertips along the surface of the agar in the second petri dish.
5. Place the top on the dish, and seal the edges with tape. Label a piece of masking tape with the letter *B*, and affix it to the bottom of the dish.
6. Use **Table 1** to record the information you will be gathering.
7. After two days, examine both petri dishes. Be aware of the development of small, white, yellow, or cream-colored dots. Each dot represents a bacterial colony. *Do not open or compromise the seal on your petri dishes.*
8. Count the number of bacterial colonies on the surface of the agar that you touched. Compare this number with the number of colonies on the surface you did not touch. If separate colonies cannot be counted, look at the percentage of the surface covered in bacteria.
9. Dispose of your petri dishes as instructed by your teacher. Do not open them. Clean the table surface with a disinfectant.

Part B. Exponential Growth of Bacteria

1. **Table 2** shows the growth of a bacterial population from a single bacteria cell.
2. Choose appropriate axes, and plot the data on a sheet of graph paper.
3. Make a best fit line to connect the dots.
4. Give your graph a title, and label the axes.

Part C. Limiting Factors

1. **Table 3** contains data gathered on the number of breeding male fur seals from 1902 to 1950.
2. Choose appropriate axes, and plot the data on a sheet of graph paper.
3. Make a best fit line to connect the dots.
4. Give your graph a title, and label the axes.

Part D. Predator-Prey Relationships

1. **Table 4** contains data on the population of snowshoe hares and lynxes during the course of 100 y.
2. Choose appropriate axes, and plot the data on a sheet of graph paper.
3. Plot the data for the hare in one color and the data for the lynx in another color.
4. Give your graph a title, and label the axes.

Part E. Human Population Growth

1. **Table 5** contains figures on the population of humans on Earth since A.D. 1.
2. Choose appropriate axes, and plot the data on a sheet of graph paper.
3. Make a best fit line to connect the dots.
4. Give your graph a title, and label the axes.

Data and Observations

Table 1

Bacterial Growth	
Petri Dish	Number of Colonies/ Percentage of Surface Covered
(A) Control	
(B) Contaminated	

Table 2

Growth of Bacteria	
Time	Number of Cells
0 min	1
20 min	2
40 min	4
60 min	8
80 min	16
100 min	32
120 min	64
240 min	4,096

Table 3

Fur Seal Population	
Year	Population
1902	1000
1911	1200
1915	3000
1917	4500
1923	3000
1924	3100
1925	3000
1932	8400
1933	8400
1936	10,700
1937	9100
1940	10,800
1942	11,000
1945	10,400
1946	11,000
1950	9500

Classic **Lab 4, How can you show a population trend?** continued

Table 4

Hare and Lynx Populations		
Year	Population	
	Snowshoe Hare	Lynx
1850	38	20
1854	90	15
1856	75	30
1857	88	32
1862	40	22
1865	30	28
1870	25	25
1872	160	40
1875	120	80
1880	40	41
1883	20	35
1885	78	33
1888	90	48
1890	87	52
1892	40	38

Table 5

Human Population	
Year	Estimated or Projected Population
A.D. 1	300,000
1200	450,000
1650	500 million
1800	1 billion
1930	2 billion
1959	3 billion
1974	4 billion
1986	5 billion
1999	6 billion
2013	7 billion
2027	8 billion

Analyze and Conclude

1. What did the bacteria consume as food in this experiment?

2. Bacteria require very little to grow and multiply. Why are there not large, visible colonies of bacteria like those in the petri dish on the things we use every day?

3. What type of curve fits the data for the growth of the bacteria in **Table 2**? How does the curve change with the rate of bacteria growth?

4. Error Analysis What were possible sources of error in your experiment?

_____

_____

5. What type of curve best fits the data for the fur seal population? Describe what happened to the population.

_____

_____

6. What happened to the fur seal population? Use the terms *limiting factor* and *carrying capacity* in your answer.

_____

_____

7. Describe the relationship you see in the graph of the snowshoe hare and the lynx population.

_____

_____

8. What type of relationship do hares and lynxes have? How would you explain their respective growth cycles?

_____

_____

9. Describe the growth of the human population in the past 350 y. Why might scientists and others be concerned about this pattern?

_____

_____

_____

Inquiry Extensions

1. Research more information on the growth of the human population. How have advances in medicine and technology contributed to the trend on the graph? Research information on death rates and birth rates of the human race. Write a short paragraph describing how technology and medicine have affected the population growth of humans.

2. What other predator-prey relationships can you think of that might exist in the wild? Research their population data, and graph their population trends. Compare the trends.

Classic
Lab 5

How do we measure biodiversity?

The number of different species in an area is an indication of the area's biodiversity. It is important to preserve biodiversity because it contributes to the balance of an ecosystem. The more diverse an ecosystem, the more stable it is. In addition, diverse ecosystems are a source of recreation and beauty.

In this lab, you will look at biomass and rainfall data from four different sites in the same ecosystem. The rainfall was consistent among the sites. The biomass at each site was measured for a period of 11 years. The biomass was determined by collecting, drying, and measuring the mass of all the plant material that could be clipped from a 0.3-m^2 area. Your job will be to analyze the data and develop a hypothesis to explain the changes in biodiversity in the communities at these sites.

Objectives
- Analyze data from four test sites.
- Infer trends in biodiversity.
- Predict which environmental factors impact biodiversity.

Materials
pen
graph paper
ruler
colored pencils
calculator

Procedure
1. Read and complete the lab safety form.
2. The data below is from four vegetation areas.
   Community 1 is native grassland.
   Community 2 was a farm 20 years ago.
   Community 3 was a farm field 31 years ago.
   Community 4 was a farm 54 years ago.
   The data is the average grams of biomass measured in each area.
3. Plot this data on a graph. Show biomass per year on the vertical axis on the left side and rainfall on the vertical axis on the right side. Use separate colors for the lines for each community and a fifth color for rainfall. Choose an appropriate scale, and label each axis.

Table 1

Annual Average Grams of Biomass				
Year	Community 1 (grams biomass per 0.3 m²)	Community 2 (grams biomass per 0.3 m²)	Community 3 (grams biomass per 0.3 m²)	Community 4 (grams biomass per 0.3 m²)	Total Annual Precipitation (cm)
1982	138	126	130	136	76.78
1983	142	123	132	145	99.24
1984	140	127	123	131	93.85
1985	138	125	135	133	80.42
1986	144	124	132	136	93.01
1987	77	5	37	56	41.30
1988	76	4	24	54	48.46
1989	112	15	38	78	58.58
1990	134	33	83	103	83.95
1991	140	56	83	105	92.43
1992	142	80	113	122	75.39

Source: David Tilman and John A. Downing, Department of Ecology, Evolution and Behavior, University of Minnesota, 1994

Data and Observations

1. Attach your graph to the space below.

Analyze and Conclude

1. What type of organisms would you expect to make up the biomass in each community?

_____

_____

2. Which community do you think is most diverse? Why?

_____

_____

_____

Classic **Lab** 5, **How do we measure biodiversity?** continued

3. What correlation do you see between the precipitation data and the biomass in each community? What does this indicate?

_____

_____

_____

_____

4. Which community had the greatest change in biomass? Which had the least? What are the reasons for this?

_____

_____

_____

_____

5. Which community recovered most rapidly after the drought? Which had the slowest recovery?

_____

_____

_____

_____

6. Error Analysis What were possible sources of error in this exercise?

_____

_____

7. Look back at your answer to question 2. How does the diversity in a community impact its biological stability?

_____

_____

_____

Inquiry Extensions

1. What other types of drastic abiotic changes could affect a community? Choose one and describe how each community in this study could be impacted and how it might rebound.

2. In terms of biodiversity, what impact could humans have on the communities described in the lab? Design a study that would trace the impact humans could have on one community. What elements would you focus on in your study? Your study can take place over a series of years.

Design Your Own
Lab 6
How much vitamin C are you getting?

Vitamin C, like other vitamins found in food, is an organic compound that serves as a helper molecule for a variety of chemical reactions in your body. Also known as ascorbic acid, vitamin C helps keep your skin and gums healthy.

Vitamin C will react with iodine and remove its amber color. If vitamin C and starch are both present, iodine will react first with the vitamin C. When the vitamin C is gone, the iodine will react with the starch, producing a blue color. This means that when the concentration of vitamin C is higher, more iodine is needed to create a color change.

Problem
Many fruit juices contain vitamin C naturally. Recently, manufacturers have started to add vitamin C to some juices and sports drinks. Select five juices, and test each one to determine which has the highest concentration of vitamin C.

Objectives
- Make predictions about the amounts of vitamin C in a variety of drinks.
- Design an experiment to compare the amounts of vitamin C in these beverages.
- Measure how different substances react in the presence of iodine.
- Draw conclusions about the nutritional value of the beverages.

Safety Precautions

WARNING: *Iodine is toxic if ingested. It will also stain hands and clothing.*

Possible Materials
50-mL beakers (6)
plastic droppers
starch solution
tincture of iodine
vitamin-C solution
orange juice
juices and sports drinks with added vitamin C (4)
additional information provided by your teacher

Hypothesis
Use what you know about vitamin C and fruit juices, and what you have been told through advertising, to write a hypothesis indicating which drink you think has the most vitamin C and why.

_____
_____
_____
_____

Plan the Experiment

1. Read and complete the lab safety form.
2. Choose five drinks to test for vitamin C. Be sure to test the drink listed in your hypothesis.
3. Decide on a procedure for testing the drinks for the amount of vitamin C. Write your procedure below. Include the materials you will use.
4. Identify the independent variable, dependent variable, constants, and control.
5. Decide how you will record your data. Design a data table to collect the information. You might want to include a column for your predicted rankings of vitamin-C content for each item.

Check the Plan

1. Make sure your teacher has approved your experimental plan before you proceed.
2. Be sure that a control is included in your experiment.
3. When you have completed the experiment, dispose of the liquids as directed by your teacher.

Record the Plan

In the space below, write your experimental procedure and make a sketch of your experimental setup.

Design Your Own **Lab** 6, **How much vitamin C are you getting?** continued

Data and Observations

1. Use the space below to create a data table of your findings, including the type of drink, the number of drops of iodine, the predicted ranking, and the actual ranking.

Analyze and Conclude

1. Which sample had the highest level of vitamin C? Which had the lowest?

_____

_____

2. Based on your observations, do drinks with natural vitamin C vary greatly from drinks with added vitamin C?

_____

_____

3. What does the vitamin-C content of the beverages imply about their nutritional value?

_____

_____

_____

4. Describe the control in your experiment. What did the control show?

_____

_____

_____

5. Error Analysis What were some possible sources of error in your experiment?

_____

_____

_____

_____

6. Compare your data to the nutritional information label provided for each drink. Do your results match the nutritional information provided? Does the information match your prediction? Compare serving sizes and the percentage of the Recommended Dietary Allowance (RDA) of vitamin C listed on the label. How much of each drink would have to be consumed in order to obtain the RDA of vitamin C?

_____

_____

_____

_____

Write and Discuss

Write a short paragraph describing your findings and indicating whether or not they support your hypothesis. Did any of the juices stand out as being higher in vitamin C than the others? Discuss any questions your results might have raised.

_____

_____

_____

_____

Inquiry Extensions

1. Packaged or processed foods are usually marked with an expiration date. You probably keep your orange juice or apple juice in a closed container in the refrigerator. Do refrigeration and a closed container help maintain the vitamin-C content? Design an experiment to determine if the vitamin-C content of orange juice changes after the expiration date or if proper storage is not available.

2. Like all nutrients, vitamin C has to survive the highly acidic environment of the stomach before it can be made available to the body. Design an experiment to determine if vitamin C is affected by stomach acid.

Lab 7

What substances or solutions act as buffers?

Organisms and individual cells must keep an internal balance, known as homeostasis. There are many external factors that can cause an individual cell or entire organism to stray from this balance. For example, many of the metabolic activities of living tissues can alter the pH in a system. Life depends on maintaining an optimal pH range. Buffers are substances that release or accept hydrogen ions to keep the pH relatively constant. Our bodies contain natural buffers. In this laboratory exercise, you will design an experiment to determine which of several living tissues has the greatest ability to buffer a solution.

Problem
Design an experiment that will test the buffering power of several animal or plant tissue solutions.

Objectives
- Form a hypothesis about the success of certain materials as buffers.
- Design an experiment to test your hypothesis.
- Control the variables in your experiments.
- Draw conclusions about the success of animal and plant solutions as buffers in biological systems.

Safety Precautions
WARNING: *The HCl and NaOH can irritate the skin.*

Possible Materials
pH meter
stirring rod
50-mL beaker
500-mL beaker
50-mL graduated cylinder
0.1M HCl
0.1M NaOH
liver solution
egg solution
gelatin solution
fruit solution
cucumber solution
buffer 7
water

Hypothesis
Use what you know about buffers and the ability of substances to act as buffers to write a hypothesis. Indicate which material will act as the best buffer and why and how this compares with the buffering ability of water.

_____

_____

_____

_____

Plan the Experiment

1. Read and complete the lab safety form.
2. Choose four solutions to test for their ability to buffer substances in a living system. Be sure to test the solutions you listed in your hypothesis.
3. Decide on a procedure for testing your solutions. In the space provided below, write your procedure for testing the solutions. Include a list of the materials you will use.
4. Identify the independent variable, dependent variable, constants, and control group.
5. Decide how you will record your data and when you will record it. Design a data table to collect information about the number of drops of HCl added and the number of drops of NaOH added to a solution as well as a space for the pH of the solutions.

Check the Plan

1. Be sure that a control group is included in your experiment and that the experimental groups vary in only one way.
2. Make sure your teacher has approved your experimental plan before you proceed.
3. When you have completed your experiment, dispose of materials as directed by your teacher.
4. Wash your hands thoroughly with soap and water.

Record the Plan

In the space below, write your experimental procedure and make a sketch of your experimental design.

Design Your Own **Lab 7, What substances or solutions act as buffers?** continued

Data and Observations

1. Use the space below to create a data table of your findings, including the pH of the solutions, the number of drops of HCl added, and the number of drops of NaOH added to the solution.

Analyze and Conclude

1. Of the solutions that you tested, which was the most effective buffer? Which was the least effective?

_____

_____

_____

2. What can you conclude about the buffering ability of water?

_____

_____

_____

3. Were plant or animal solutions the best buffers? Why do you think this is the case? What evidence do you have about the effectiveness of plant and animal buffers? Explain.

_____

_____

_____

4. Describe the control in your experiment. What did the control show?

_____

_____

_____

5. Error Analysis What were some possible sources of error in your experiment?

_____

_____

_____

6. Exchange your procedure and data with another group in your class for peer review. What do their data indicate about the ability of different substances to be buffers?

_____

_____

_____

Write and Discuss

Write a short paragraph describing your findings and indicating whether or not they support your hypothesis. Discuss any questions your findings might have raised.

_____

_____

_____

_____

Inquiry Extensions

1. Human blood normally has a pH of about 7.4. What activities or conditions can change blood pH? Is it easy or difficult to change blood pH? Research diseases or conditions that can change the pH. Describe the symptoms and possible outcomes, and discuss the remedies for such issues. Report your findings to the class in the form of a public service announcement.

2. What are the buffering mechanisms used by the body to maintain proper pH? How do they work?

Classic
Lab 8

Why do cells divide?

When cells grow to a certain size, their rate of growth slows until they stop growing. At this point, they have reached their size limit. A cell that has reached its size limit divides into two smaller cells. In this lab, you will explore one of the factors that limit cell size: the relationship between the size of the cell—specifically, its surface area and volume—and how efficiently substances diffuse across its cell membrane.

Objectives

- Model cells of different sizes with agar cubes.
- Model the diffusion of materials across a cell membrane.
- Calculate the surface area-to-volume ratio for model cells.
- Form a hypothesis about how cell division affects a cell's ability to absorb materials.

Materials

agar
beaker
timer
calculator
plastic ruler
100 mL 0.1*M* solution of hydrochloric acid
kitchen knife
plastic spoons
paper towels

Safety Precautions

WARNING: *Use caution when handling hydrochloric acid.*

Procedure

Part A. Setting Up the Experiment

1. Read and complete the lab safety form.
2. Obtain a block of agar containing phenolphthalein from your teacher. Recall that phenolphthalein turns pink in the presence of a base. It will become colorless in an acid.
3. Use a ruler to measure and a kitchen knife to cut three blocks out of the agar. One should be 3 cm on each side, one should be 2 cm on each side, and one should be 1 cm on each side.
4. **Figure 1** Place the three agar cubes inside the beaker. Cover with 100 mL dilute hydrochloric acid solution.

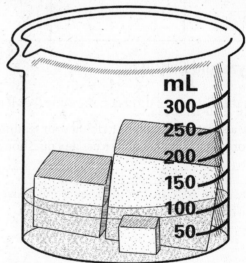

Figure 1

5. Leave the agar blocks in the dilute hydrochloric acid for a total of 10 min. Use a spoon to turn them every few minutes to ensure that they are soaking evenly.
6. Complete the data table on the next page.

Classic **Lab 8, Why do cells divide?** continued

Part B. Measuring Diffusion

1. After 10 min, carefully use the plastic spoons to remove the agar blocks. Blot them dry with paper towels. Use care not to splash HCl on skin; it will cause burns.

2. Use the edge of the plastic ruler to cut each block in half. Measure the depth of the uncolored area in centimeters, recording the measurement to the nearest millimeter. This shows the depth of diffusion. Record these values in **Table 1**.

3. Complete **Table 1**, and answer the questions that follow.

4. You might need the following formulas:

   surface area = length × width × number of surfaces

   volume of a cube = length × width × height

   Use a calculator for your calculations if necessary.

5. Wash your hands with soap and water, and dispose of the materials as instructed by your teacher.

Data and Observations

Table 1

Agar Data				
Cube Size	Surface Area	Volume	Ratio	Depth of Diffusion
3 cm/side				
2 cm/side				
1 cm/side				

Analyze and Conclude

1. Is the distance of diffusion the same for all of the blocks? Explain.

   _____

   _____

2. Based on your answer to the question above, do you think that the depth of diffusion is the same in all cells? Explain.

   _____

   _____

3. List the agar cubes in order of size, from largest to smallest. Then list them in order of surface area-to-volume ratio (from largest ratio to smallest ratio). How do these lists compare?

   _____

   _____

   _____

Classic **Lab 8, Why do cells divide?** continued

4. Suppose you were given a microscopic, cube-shaped onion cell that was 0.01 cm/side. What would be the surface area-to-volume ratio of that cube?

_____

_____

5. Which block has the greatest surface area-to-volume ratio—the onion cube or the 3 cm/side cube you used in this lab?

_____

_____

6. What is the relationship between surface area-to-volume ratio and diffusion across a cell?

_____

_____

7. What happens to diffusion as a cell grows?

_____

_____

_____

8. **Error Analysis** What are some possible sources of error in your experiment?

_____

_____

_____

9. Form a hypothesis to explain how cell division affects a cell's ability to absorb the material necessary for growth. Base your answer on your observations of the surface area-to-volume ratio.

_____

_____

_____

Inquiry Extensions

1. Which cells in the human body divide most frequently? Why is this? What activities or conditions spur cell division? What slows it down?

2. During adolescence the human body grows at a rate faster than at any other time after infancy. Explain how what you learned in this lab plays out in the human body during adolescence.

Design Your Own
Lab 9

How many calories do different foods contain?

As you know, energy is stored in the bonds of large molecules. This means that chemicals in gasoline, natural gas, and food are forms of stored energy. Molecules of oxygen can break up the large molecules, releasing the energy. If this release of energy occurs quickly, it is called combustion (like the process that occurs in a car engine).

In your cells, the burning of food (glucose) occurs more slowly, which means that the energy is released more slowly. Some athletes, such as long-distance runners, take advantage of this phenomenon by eating large amounts of pasta the night before a race. The energy from the carbohydrate-rich food usually takes about a day to be released. Enzymes guide every step of this cellular respiration process so that the reaction does not happen all at the same time in an explosive fashion.

Energy is measured in calories. One calorie is the amount of energy needed to raise 1 mL of water 1° C. The measurement used on nutrition labels is Calorie, which equals 1000 calories. Calculate the number of calories using the following formula:

Number of calories = number of degrees the temperature rises × volume of water

Calculate the number of calories per gram using the following formula:

Number of calories ÷ gram = number of calories ÷ mass change

Problem
Determine the amount of calories per gram contained in different types of food. Identify which foods provide the most energy.

Objectives
- Identify foods to test.
- Assemble a simple calorimeter.
- Design an experiment to test a hypothesis.
- Draw conclusions about available energy in food.

Safety Precautions

WARNING: *Use care when touching hot metal. Be careful with pins and dissecting probe—they can pierce the skin.*

Possible Materials
metal coffee can (clean)
metal soup can (clean)
cork
straight pins or dissecting pins
aluminum foil
test-tube holder
graduated cylinder
masking tape
scale
food samples such as rice cakes, peanuts,
    dried beans, dried cheese, marshmallows
temperature probe or thermometer (non-mercury)
matches
candles
locking tongs
dissecting probe
weigh boats

Hypothesis
Use what you know about fats, calories, and foods in general to write a hypothesis indicating which types of foods have the greatest amounts of calories.

_____

_____

_____

_____

Plan the Experiment

1. Read and complete the lab safety form.
2. Choose two different foods to test. Be sure to test the food you mentioned in your hypothesis.
3. Decide on a procedure for determining the calories in the food. In the space provided, write the procedure, including a list of materials you will use.
4. Identify the independent and dependent variables, the constants, and the control group in your experiment.
5. Decide how you will record your data. Consider gathering temperature information every 20 s. Design a data table to collect information about the beginning weight, end weight, water volume, and changes in temperature. Leave space in your data table for the calories.

Check the Plan

1. Be sure that a control group is included in your experiment and that the experimental group varies in only one way.
2. Make sure your teacher has approved your design and your experimental plan before you proceed.
3. When you have completed the experiment, dispose of the materials as instructed by your teacher.

Record the Plan

In the space below, write your experimental procedure and make a sketch of your experimental setup.

Design Your Own **Lab 9, How many calories do different foods contain?** continued

Data and Observations

1. Use the space below to create a data table of your findings.

Analyze and Conclude

1. Which food had the highest number of calories? Which food had the least?

_____

_____

_____

2. What does the data tell you about the energy the body can get from each of these foods? Explain.

_____

_____

_____

3. Why do you think there were differences in the calories of the foods?

_____

_____

_____

4. Use what you learned to write a definition for cellular energy, in your own words.

_____

_____

_____

5. Did all the food energy go into the water? Explain.

_____

_____

_____

6. Error Analysis What were some possible sources of error in your experiment?

_____

_____

_____

_____

7. What is the primary difference between the form in which most plants store energy and the form in which animals store energy? How does each method benefit that type of organism?

_____

_____

_____

_____

Write and Discuss

Write a short paragraph describing your findings and indicating whether or not they support your hypothesis. Discuss any questions your results might have raised.

_____

_____

_____

_____

Inquiry Extensions

1. Some cells are able to obtain energy from food without using oxygen. These organisms obtain their energy through fermentation, an energy-releasing process that does not require oxygen. Yeast is one such organism. Design an experiment to examine one factor that might impact the rate of fermentation.

2. Inorganic and organic fuels also release energy when burned. Design an experiment that will compare the amount of energy released by several types of fuel. For example, you might want to compare the calories that are released when burning different types of wood (hardwood v. soft wood), or the amount of energy released with the burning of different types of coal (lignite, bituminous, anthracite). When you have completed your experiment, draft a declaration on the environmental impact these fuels will have.

Design Your Own
Lab 10

What can affect the rate of photosynthesis?

Green plants can turn inorganic chemicals into organic food (stored energy). By synthesizing macromolecules, photosynthetic organisms transform nonliving materials into the building blocks of life. Plants take in water and carbon dioxide and make food through the process of photosynthesis. Light energy and chlorophyll are needed for this conversion. The amount of light a plant receives changes on a daily, weekly, and monthly basis.

   Oxygen is a by-product of photosynthesis. The change in light intensity will probably impact the amount of oxygen a plant will produce. In this laboratory exercise, you will design an experiment that examines how light intensity impacts the rate of photosynthesis.

Problem
Determine how the intensity of light impacts the rate of photosynthesis.

Objectives
- Formulate a hypothesis about the connection between light intensity and oxygen production in photosynthesis.
- Design an experiment to test this hypothesis.
- Control variables, and use a control during your experiments.
- Draw a conclusion about the rate of photosynthesis.

Safety Precautions

Possible Materials
large glass jars (3)
aged tap water
baking soda (sodium bicarbonate)
scale
Elodea samples
ruler
scissors
small glass funnels (3)
test tubes (3)
lamp
medium to large box lined with white paper
medium to large box lined with gray paper

Hypothesis
Use what you know about photosynthesis to develop a hypothesis about the impact light intensity can have on the rate of photosynthesis.

_____

_____

_____

_____

Plan the Experiment

1. Read and complete the lab safety form.
2. Make a list of the impacts light might have on the rate of photosynthesis. Be sure to include the impacts you listed in your hypothesis.
3. Decide on a procedure for testing your hypothesis. In the space provided, write your procedure. Include a list of the materials you will use.
4. Identify the independent variable, dependent variable, constants, and control group.
5. Decide how you will record your data and when you will record it. Design a data table to collect information about the amount of oxygen produced.

Check the Plan

1. Be sure that a control group is included in your experiment and that the experimental groups vary in only one way.
2. Make sure your teacher has approved your experimental plan before you proceed.
3. When you have completed your experiment, dispose of materials as directed by your teacher.

Record the Plan

In the space below, write your experimental procedure and make a sketch of your experimental setup.

Design Your Own **Lab** 10, **What can affect the rate of photosynthesis?** continued

Data and Observations

1. Use the space below to create a data table of your findings, including information about the amount of oxygen produced.

Analyze and Conclude

1. What evidence do you have that plants need light for photosynthesis?

_____

_____

2. Is oxygen given off during photosynthesis? What is your evidence?

_____

_____

_____

3. What purpose did the sodium bicarbonate serve?

_____

_____

_____

4. What impact does light intensity have on photosynthesis? Make a general statement that explains your results.

_____

_____

_____

5. Describe the control in your experiment. What did your control show?

_____

_____

6. **Error Analysis** What were some possible sources of error in your experiment?

_____

_____

_____

7. Exchange your procedure and data with another group in your class for peer review. What do their data indicate about the impact light intensity has on photosynthesis?

_____

_____

_____

Write and Discuss

Write a short paragraph describing your findings and indicating whether or not they support your hypothesis. Discuss any questions your results might have raised.

_____

_____

_____

_____

Inquiry Extensions

1. Does the daily cycle of day and night impact a plant's overall rate of photosynthesis? Design an experiment that examines the changes in the release of oxygen between day and night. Then discuss how house plants or conifers might respond to seasonal changes and changes in the length of the day over the course of a year.

2. Why is it important to know about the production of oxygen by plants? What would happen to life on Earth if all plants disappeared? Now that you have seen how much oxygen a single plant can produce in a single day, research the total amount of oxygen in the atmosphere, the amount of oxygen needed to support life as we know it, and the approximate percentage of plants v. animal life on Earth. Use this information to create a time line that shows what would happen if there were no plants. Illustrate your time line with diagrams, and predict what changes would occur.

Classic

Lab 11

How long does each phase of the cell cycle last?

Have you ever considered what happens to you when you have an injury or you are in the middle of a growth spurt? What exactly is going on at the cellular level? Whether you are injured or you are growing, cells are busy growing and dividing during their cell cycle. During this investigation, you will be exploring each phase of the cell cycle by asking questions such as "What happens in each phase?" and "How long does each phase last?"

The cell cycle has a series of phases: interphase (which includes two growth phases and a DNA synthesis phase), mitosis, and cytokinesis. Mitosis can be broken into four different stages: prophase, metaphase, anaphase, and telophase. Each of these phases takes a different amount of time.

In this lab, you will be examining onion root cells under a microscope. You will find that different cells of the onion are at different stages in the cell cycle. Your job will be to count the number of cells representing each phase of the cell cycle. The cell cycle for onion root tips is about 24 h (or 1440 min). You will use the number of cells engaged in each phase as an indicator of how much time the cell spends in that phase.

Objectives

- Use a microscope to identify cells in an onion root tip.
- Identify the different stages of the cell cycle in onion cells.
- Count the number of cells in each stage of the cell cycle.
- Calculate the amount of time cells spend in each stage of the cell cycle.

Materials

microscope
colored pencils
calculator
prepared slide of onion root tip cells undergoing cell division

Safety Precautions

Classic **Lab** 11, **How long does each phase of the cell cycle last?** continued

Procedure

1. Read and complete the lab safety form.
2. Familiarize yourself with the stages of the cell cycle. Sketch out the phases of cell division to help you identify those stages when you see them under a microscope.
3. Work with a partner and set up your microscope. One partner will act as the Observer and use the microscope to locate onion cells. The second partner will act as the Recorder and will tally the stages as the Observer calls them out.
4. Obtain a prepared onion root tip slide from your teacher and focus on it under low power.
5. Wait for instructions from your teacher. You will be timed during your observation of this onion cell.
6. Switch to high power and locate the region of active growth, just above the root cap.
7. The Observer should start with one long column of cells on the left side of the field of view. Identify the stage of mitosis that the cell is in. Call out the stage to your partner. Complete five to seven columns of cells. Then switch jobs with your partner.

8. The Recorder uses tally marks (in sets of five) to record the stages in **Table 1** as the partner calls them out.
9. Total the number of cells you find of each type. Put that number in the *Your Total* column of **Table 1**.
10. Wait until all your classmates have finished, and place their data (including your own) in the *Class Total* column of **Table 1**.
11. Calculate the percentage of each stage. Record this information in **Table 1**.
12. Assuming it takes 24 h for a cell to complete the cell cycle, calculate how long each stage takes (in hours). Hint: You will need to use your percentages for this calculation. Record your answers in **Table 1**.

Data and Observations

Table 1

Cell Cycle Data							
Stage	Description	Tally Marks	Your Total	Class Total	Total	Percent of Total	Time of Stage
Interphase							
Prophase							
Metaphase							
Anaphase							
Telophase							

Classic **Lab 11, How long does each phase of the cell cycle last?** continued

Data and Observations continued

1. In the space below, sketch and label an example of each stage of the cell cycle you observed.

Analyze and Conclude

1. Which stage of the cell cycle did you observe most often?

_____

_____

_____

2. What process must take place before mitosis can begin?

_____

_____

_____

3. Why might each stage of mitosis last a different amount of time? Explain.

_____

_____

_____

4. What can you infer about the relative length of time that each stage lasts?

_____

_____

_____

5. What marks the completion of telophase? Describe any structures that you saw that indicate the end of that phase.

_____

_____

_____

6. Error Analysis What are possible sources of error in your experiment?

_____

_____

_____

7. Explain how the cell cycle can be described as multiplying by dividing.

_____

_____

_____

Inquiry Extensions

1. Interphase and mitosis are similar in plant and animal cells, except that the centrioles appear during prophase in animal cells. Predict whether animal cells or plant cells spend a longer time in mitosis. Design an experiment to test your prediction.

2. Why is it important for you to think about mitosis and consider the amount of time cells spend in each phase? What does the cell cycle have to do with your life?

Design Your Own Lab 12

Green or yellow?

The phenotype of an individual organism depends on both the environment and genes. A tree's genotype does not change, but the size, shape, and greenness of the leaves will vary depending on the amount of sunlight and wind. Environmental factors such as temperature, nutrition, exercise, and exposure to sunlight can influence the phenotype of an organism.

The product of a genotype is generally not a single, rigidly defined phenotype, but a range of possibilities influenced by the environment. For instance, most people have muscles (as determined by the genotype), but people who lift heavy weights often show larger, stronger, and better-defined muscles (the expression of the phenotype) than people who do not do strength training.

Many plants that are sold commercially come with tags specifying the best light conditions for healthy growth and the best display of foliage or flowers. This is an example of knowing what environmental conditions will affect the expression of phenotypes.

Problem

Design an experiment to examine the impact that exposure to sunlight has on the phenotypes of flowering tobacco seedlings.

Objectives

- Develop a hypothesis to predict the impact sunlight will have on the phenotypes of flowering tobacco seedlings.
- Design an experiment to test this hypothesis.
- Identify a control in the experiment.
- Draw conclusions from the plants that grow.

Safety Precautions

WARNING: *Be aware of mold growth on wet filter paper or seeds. Discard if mold is present.*
Be sure to wipe up any water that spills on the floor to prevent slips and falls.

Possible Materials

Nicotiana alata (flowering tobacco) seeds
filter paper
petri dishes
fine-point permanent marker
water
sunny spot in the room
dark corner in the room
metric ruler

Hypothesis

Use what you know about genetics to develop a hypothesis that will explain how sunlight impacts the phenotypes of a *Nicotiana alata* (flowering tobacco) plant, including the height or color of the seedlings, or number and quality of their leaves.

_____

_____

_____

_____

Plan the Experiment

1. Read and complete the lab safety form.
2. Choose the locations and the conditions in which you want to grow your *Nicotiana* seedlings.
3. Identify the independent variable, dependent variable, constants, and control group.
4. Decide how you will record your data and when you will record it. Design a table to collect information about the location of your samples and the growth, or lack of growth, that you observe.

Check the Plan

1. Make sure your teacher has approved your experimental plan before you proceed.
2. Be sure that a control group is included in your experiment and that the experimental groups vary in only one way.
3. Observe the growth of the seedlings in the different locations.
4. When you have completed the experiment, dispose of your seedlings as directed by your teacher.

Record the Plan

In the space provided, write your experimental procedure and make a sketch of the experimental setup.

Design Your Own **Lab** 12, **Green or yellow?** continued

Data and Observations

1. In the space below, make detailed drawings of your *Nicotiana* seeds as they grow. Be sure to make quantitative measurements.

2. Use the space below to create a data table for your findings, including the location of the seeds and the resulting color of the seedlings.

Analyze and Conclude

1. Of the seedlings that were grown in the dark, what was the ratio of green to yellow? What was the ratio of green to yellow for plants that were grown in the sunlight?

   _____

   _____

   _____

2. Based on your observations, does environment have an impact on the phenotype of these plants? Which environmental factor in particular?

   _____

   _____

   _____

3. Why do you think there were differences?

_____

_____

_____

4. Describe the control in your experiment. What did the control show?

_____

_____

_____

5. Error Analysis What were some possible sources of error in your experiment?

_____

_____

_____

6. Exchange your procedure and data with another group in your class for peer review. What do their data indicate about the seedlings' phenotypes?

_____

_____

_____

Write and Discuss

Write a short paragraph describing your findings and indicating whether or not they support your hypothesis. Discuss any questions your results might have raised.

_____

_____

_____

_____

_____

Inquiry Extensions

1. Do other environmental conditions influence the phenotypes of tobacco seedlings? Choose another environmental factor to study, and design an experiment to test if that factor influences the phenotype.

2. Did you notice any differences between your experimental group and the control group that you did not anticipate? Make further comparisons between the two groups, and assemble the quantitative data. Design an experiment or conduct research to explore some of those unanticipated outcomes.

What are the chances?

Genetic disorders are abnormal conditions that are inherited through genes or chromosomes. Some genetic disorders are caused by mutations in the DNA of genes. Others are caused by changes in the overall structure or number of chromosomes.

Cystic fibrosis is a genetic disorder in which the body produces unusually thick mucus in the lungs and intestines. This makes it difficult for a person with cystic fibrosis to breathe or digest food. Cystic fibrosis is caused by a recessive allele. At this time, the symptoms of cystic fibrosis can be controlled, but there is no cure for this disease.

In this lab, you will determine the probability that cystic fibrosis will appear in the children of a couple who carry the traits for the disease.

Objectives
- Construct a pedigree for a family.
- Determine the probability of a couple having a child with the genetic disorder.

Materials
index cards (two colors—blue and pink)
scissors
pencil

Safety Precautions

WARNING: *Be careful when using scissors—they are sharp and can pierce or cut the skin.*

Procedure
1. Read and complete the lab safety form.
2. Read the following family history:
   Anthony and Emma have a daughter named Kathryn. Kathryn has been diagnosed with cystic fibrosis. Anthony and Emma are both healthy. Anthony's parents are both healthy. Emma's parents are both healthy. Anthony has a brother, named Corbin, who has cystic fibrosis.
3. In the space in the *Data and Observations* section, draw a pedigree that shows all the family members mentioned here. Use circles to represent the females and squares to represent the males. Shade the circles or squares representing the people who currently have cystic fibrosis.

4. Use the index cards to create a set of cards to represent the alleles. Cut three index cards of each color into fourths. On the 12 blue cards, write *F* to represent the dominant normal allele. On the 12 pink cards, write *f* for the recessive allele.
5. Use the cards to represent Kathryn's alleles. Write her genotype next to the pedigree symbol for Kathryn.
6. Use the cards to show Corbin's alleles and write the genotype next to his symbol.
7. Use the cards to determine what genotypes Anthony and Emma must have. Write their genotypes next to their pedigree symbol.
8. Use the index cards to determine the genotypes of the other family members. Fill in each person's genotype next to their symbol in the pedigree. Write in all possible genotypes.

Data and Observations

1. Use this space to draw a pedigree of Anthony and Emma's family.

Classic **Lab** 13, **What are the chances?** continued

Analyze and Conclude

1. What were the genotypes of Anthony's parents? What were the genotypes of Emma's parents?

   _____

   _____

   _____

2. Anthony also has a sister, Zoë. What is the probability that she has cystic fibrosis? Explain.

   _____

   _____

   _____

   _____

3. What is the probability that Anthony and Emma will have another child with cystic fibrosis?

   _____

   _____

   _____

   _____

4. Why is information about several generations of family members necessary to get a good idea about a hereditary condition? Explain.

   _____

   _____

   _____

   _____

5. Do you think cystic fibrosis is a sex-linked genetic disorder? Explain.

   _____

   _____

   _____

   _____

Classic **Lab** 13, **What are the chances?** continued

6. Error Analysis What are some possible sources of error in an exercise like this one?

_____

_____

_____

_____

_____

Inquiry Extensions

1. Some genetic disorders are more common in certain ethnic groups. Select an ethnic group, research the prevalence of inherited diseases for that group, and create a fictional pedigree to illustrate transmission patterns. Provide additional information on the prevalence of the disease in other ethnic groups.

2. Some diseases or traits are sex-linked. Research one such disease or trait, and report on how it is transmitted and which individuals are affected. In addition, make a fictional pedigree to demonstrate how a sex-linked disease would appear in a family across generations.

Classic
Lab 14

What is DNA?

There are 6 billion bits of information coded by DNA in each of our nucleated cells (a bit is a measure of information). Each human cell contains 21 times the information that is found in a set of encyclopedias, which is thought to have about 280 million letters.

DNA can be extracted from any living thing. Deoxyribonucleic acid is the blueprint for life, and all living things contain DNA. In this lab, you will conduct an experiment to extract DNA from different sources or to extract it from the same source by using different protocols. You will then compare your yield of DNA with that of your classmates.

In this procedure, mechanical mashing helps break down the cell walls. Heating destroys enzymes that can shear the DNA into small pieces. The detergent dissolves the lipids in the cell membranes and nuclear envelope (just like grease is dissolved in soapy dishwater). Once the membranes are dissolved, the DNA is free and very soluble in water. The enzyme (meat tenderizer) cleans the proteins, which might cling to the DNA. When the alcohol is added, the DNA clumps together and precipitates at the water/ethanol interface because DNA is not soluble in ethanol.

Objectives

- Extract DNA from organic sources.
- Compare the amount of DNA yielded from different sources.
- Design experiments to compare different extraction protocols.

Materials

various sources of DNA
various types of alcohol
various detergents
various sources of enzymes
non-iodized salt
hot-water bath
ice bath
ice cubes
beakers (50-mL, 250-mL, or large test tubes)
paper plate
knife and fork
10-mL graduated cylinder
cheesecloth
funnel
stirring rod
blender
balance
filter paper
wire inoculating loop
thermometer

Safety Precautions

WARNING: *Keep the cover on the blender when it is in operation. Use caution with the hot water—it can burn. Use caution with alcohol—it is flammable and can irritate the skin.*

Procedure
Part A. Extracting the DNA

1. Read and complete the lab safety form.

2. Working in two teams of two, decide whether you will test different sources of DNA or if you will vary your extraction protocol by manipulating the variables in bold type below. Design your experiment so that one team is the control. Your teacher will provide you with options. Decide your question and hypothesis for the experiment. Record these in the *Data and Observations* section. Record your design criteria in **Table 1**.

3. Prepare 10 mL of chilled **ethyl alcohol**.

4. Prepare the DNA extraction solution. Dissolve 2 g of salt in 90 mL of water in a 250-mL beaker. Then add 10 mL of **detergent**. Stir gently so you do not generate many suds.

5. Prepare the **source of DNA**. Place the DNA source on a paper plate or piece of waxed paper. Use a fork (and knife if needed) to mash it thoroughly. Place 30 g of the mashed DNA source in the extraction solution.

6. Check the **temperature** of your hot-water bath; the ideal temperature is 60°C. Adjust as needed by adding ice cubes or increasing the heat source. Place the beaker with the DNA source and extraction solution into the hot-water bath. Incubate the solution for **12 min**. Stir occasionally to distribute the heat. The temperature of the bath must be maintained during this incubation period.

7. After the incubation period, **transfer the solution to a blender and pulse the blender three to five times.** Do not let the blender run—the solution will become too sudsy. Return the solution to the beaker, and place the blended material into an ice bath for 5 min.

8. Pour the cooled extraction solution into your filter system. (Filter setups can include a funnel and cheesecloth or similar materials.) Allow the liquid to filter until you have 20 mL of filtrate in the bottom of a 50-mL beaker or large test tube.

9. Add a pinch of **enzyme,** and gently stir.

Part B. Precipitating and Drying the DNA

1. Tilt the beaker or test tube, and slowly add the 10 mL of cold alcohol, allowing it to slide down the side of the vessel and form a layer on top of the extraction solution.

2. Place the extraction/alcohol solution so that you can observe what happens where the alcohol and filtrate layers meet. Record your observations.

3. Let the solution sit for 2 min without disturbing it. A white precipitate will form in the alcohol layer. This is DNA, and it will appear as slimy, white strands or clumps of material.

4. Set a piece of filter paper on the balance, and record its mass in **Table 1**.

5. Use a wire loop to collect all the DNA from the top alcohol layer. Place the DNA you collect on the filter paper, and spread it out as much as possible; it will dry more slowly if it is clumped.

6. Clean up your lab materials as directed by your teacher. Wash your hands with soap and water when you are done.

7. Let the DNA sit for 24 h or until it is absolutely dry. Calculate the mass of the DNA collected [(Mass of the filter paper + DNA) – (Mass of the filter paper before DNA) = Mass of the DNA].

Classic **Lab** 14, **What is DNA?** continued

Data and Observations

1. What question are you exploring?

_____

_____

_____

_____

2. What is your hypothesis?

_____

_____

_____

Table 1

Source of DNA or Change in Protocol		Mass of Filter Paper	Mass of DNA + Filter Paper	Mass of DNA
Independent variable				
Control				

Analyze and Conclude

1. We cannot isolate most of the other molecules that make up living things as easily as we can the DNA. Why do you think this is so?

_____

_____

_____

2. Did the DNA you collected come out in clumps or strands? Explain why this might be.

_____

_____

_____

Classic **Lab 14, What is DNA?** continued

3. Calculate how many times to the Moon and back a human's DNA would reach if it were removed from each cell and each strand were laid end to end. Each human cell nucleus holds about 2 m of DNA, and a typical adult human is composed of 60 trillion cells. The distance from Earth to the Moon is 380,000 km.

4. As part of the procedure, you denatured proteins. What are the two major roles of proteins in living things?

5. What properties does the detergent possess that make this experiment possible?

6. Think about the amount of material with which you started. Now think about the amount of material that might typically be found at a crime scene (for example, a single hair follicle or dried saliva on an envelope). How would the DNA extraction process change if you were working with such a small sample?

7. **Error Analysis** What were possible sources of error in your procedure?

Inquiry Extensions

1. Do different sources give different amounts of DNA? Do certain properties of substances make it possible to extract more DNA? Design an experiment to test your hypotheses.
2. Does the type of detergent used make a difference in the success of extracting DNA? Do powdered soaps work as well as liquid? What about shampoo? Design an experiment to test your hypotheses.

Classic
Lab 15

Who did it?

An individual's DNA forms a unique pattern of bands that can be used to identify the person. This unique banding pattern produced by fragments of your DNA is called DNA fingerprinting. (Interestingly, identical twins have unique fingerprints, but their DNA markers are exactly the same.) Genetic markers can help identify the differences between two DNA samples. Genetic markers are specific stretches of DNA that vary among individuals.

Scientists use polymerase chain reactions (PCR) and gel electrophoresis to make a DNA fingerprint. PCR allows many copies of a certain segment of DNA to be made without using living cells. PCR can make multiple copies of one particular segment from within a large length of DNA. The DNA fragments are then separated by gel electrophoresis. In this process, the molecules or portions of molecules are separated according to length.

These two techniques, when used together, allow law enforcement to gather the smallest amount of evidence at a crime scene. DNA gathered from a blood or hair sample at a crime scene can then be compared to DNA from a suspect's blood or saliva.

Objectives
- Use models to represent DNA fingerprints.
- Infer why DNA patterns differ between individuals.
- Draw conclusions about which suspect was present at a crime scene.

Materials
mock DNA fingerprint set
magnifying lens
ruler
Figure 1

Procedure
Part A. Identify a Pattern

1. Read and complete the lab safety form.
2. Obtain a set of suspect "DNA fingerprints" from your teacher. These might look familiar. They are the Universal Product Codes (UPC) from several common products. For the purpose of this lab, these bar codes will be used to model DNA fingerprints.
3. Also obtain an envelope labeled *Crime Scene Data*. Your job will be to determine which of the suspects left behind evidence at a crime scene.
4. Use a magnifying lens to examine the DNA fingerprints carefully. The suspect whose DNA fingerprint matches the sample from the crime lab will be the suspect who is arrested for the crime.
5. Once you have isolated a suspect, take your answer to your teacher for review. If you are correct, you will be able to move ahead to the next section.

Part B. Using DNA Evidence

1. **Figure 1** is an example of DNA patterns made by gel electrophoresis derived from a DNA database. A series of bank robberies took place in one town during the past four days. Your job will be to determine if the robberies were linked and if any of the suspects now in custody are the guilty parties.

2. The First National Bank was robbed at noon on Monday. The bank robber ran up to the drive-through window and demanded money. The robber made off with an unspecified amount of money but cut a finger smashing the surveillance camera. Police detectives analyzed the blood sample. The DNA sample is the sample shown in the first column in **Figure 1.** It is labeled *Bank 1.*

3. The Second National Bank was robbed at 11 A.M. on Tuesday. This time, the bank robber entered the bank and handed the teller a note for an unspecified amount of money. The teller handed over the money but kept the envelope. Luckily for detectives, the robber licked the envelope and left behind a DNA sample. This sample is in the second column of **Figure 1.** It is labeled *Bank 2.*

4. The Third National Bank was robbed at 10 A.M. on Wednesday. The robber demanded money and left without leaving any evidence at the teller's station. However, the robber had been chewing gum and, just before stepping up to the teller, dropped the gum in the trash barrel. Observant bank patrons alerted the police, and the gum was collected and analyzed. DNA was extracted, and this DNA sample is seen in the third column of **Figure 1.** It is labeled *Bank 3.*

5. The police have found three possible suspects, and the suspects have consented to give DNA samples.

6. Examine the DNA strands and see if any of the suspects should be arrested for the crime or crimes.

7. Record your conclusions in **Table 1,** and answer the questions that follow. Indicate in the table whether or not the DNA from each suspect was found at a crime scene. Use an X to indicate a positive match.

DNA Gel Electrophoresis

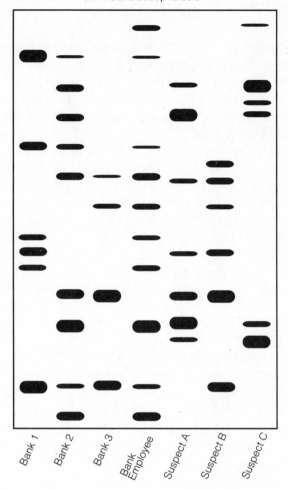

Figure 1

Classic **Lab** 15, **Who did it?** continued

Data and Observations
Table 1

DNA Data			
	Bank 1	**Bank 2**	**Bank 3**
Suspect A			
Suspect B			
Suspect C			

Analyze and Conclude

1. Were any of the three bank robberies committed by the same person? Explain how you know.

_____

_____

_____

_____

2. Is there a definite suspect for each bank robbery? Should any of the suspects be released? Do detectives need to gather more evidence in any of the cases? Explain.

_____

_____

_____

_____

3. Suppose detectives learn that Suspect A has an identical twin. How will this change their investigation?

_____

_____

_____

_____

_____

Classic **Lab 15, Who did it?** continued

4. Error Analysis What sort of errors can occur when collecting and examining DNA samples?

_____

_____

_____

_____

5. How did the exercise where you examined UPC symbols compare with examining DNA fingerprints? How were the experiences similar or different?

_____

_____

_____

_____

6. Each suspect had a very different DNA fingerprint. Why do DNA fingerprints vary so much from person to person?

_____

_____

_____

_____

Inquiry Extensions

1. Suppose Suspect B is arrested for one of the bank robberies. You are his or her defense lawyer. Write a short paragraph stating why you plan to use the DNA evidence to help get your client acquitted.
2. Many criminal cases are solved using DNA evidence. Conduct research for a recent case which has been reopened or solved using DNA evidence. Report your findings to your classmates.

Classic Lab 16

How do species compare?

In the 1740s, a French scientist first proposed the idea that all organisms are descended from a single common ancestor. **Figure 1** shows a widely accepted phylogenetic tree (an organizational chart) showing the relationships of major groups of animals. (The arrowhead on the left side of the chart points back toward the presumptive single ancestor.)

Scientists who have studied protein sequences have found evidence to support the idea of a common ancestor. They have determined that when two species share a similar sequence of protein chains, the species must have shared a common ancestor. The closer the sequences are to each other, the more recently the two species shared an ancestor. Numerous differences in the sequences indicate the two species are not as closely related.

For example, the amino acid sequence in the protein cytochrome c of humans exactly matches the sequence in chimpanzees. The human sequence differs only by one position when compared to that of a rhesus monkey. But for animals that we are clearly not closely related to, the sequence shows an even greater variance. The amino acid sequence of humans differs from that of a chicken by 18 positions and from a turtle's by 19 positions.

In this investigation, you will look at the amino acid sequences from a variety of organisms and discuss how closely related they are.

Objectives
- Examine a table of amino acid data.
- Interpret the table and find relationships.
- Draw conclusions about how closely related species are.

Material
copy of amino acid table

Procedure
Part A. Predict Relationships
1. Read and complete the lab safety form.
2. You will be using **Table 1** to examine amino acid sequences from the following animals: horses, donkeys, rabbits, snakes, turtles, and whales.
3. **Table 1** shows only a small segment of the sequence of amino acids within the cytochrome c protein. There are 104 amino acids in this protein. **Table 1** shows the sequence between positions 39 and 53.

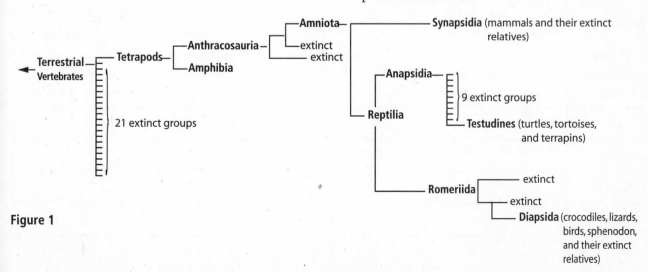

Figure 1

4. Use the characteristics of the animals, such as appearance, habitat, or diet, for example, to make a prediction as to which of them are most closely related. List these in the *Data and Observation* section of this lab.

5. Indicate which animals you think are the least closely related.

Part B. Examine Amino Acid Sequences

1. Compare the amino acid sequence of the horse to the rest of the animals. How many amino acids differ between the species? Record your results in the second column of **Table 2**.

2. Study the relationships between the species indicated in the first column of **Table 2**. What species shares the most similar sequence with a horse? Which species have sequences that are dissimilar?

Table 1

Position of the Amino Acids in Cytochrome c															
Position of the amino acid	39	40	41	42	43	44	45	46	47	48	49	50	51	52	53
Horse	A	B	C	D	E	F	G	H	I	J	K	L	M	N	O
Whale	A	B	C	D	E	Y	G	H	Z	J	K	L	M	N	O
Turtle	A	B	C	D	E	V	G	H	Z	J	K	U	M	N	O
Rabbit	A	B	C	D	E	Y	G	H	Z	J	K	L	M	N	O
Donkey	A	B	C	D	E	F	G	H	Z	J	K	L	M	N	O
Snake	A	B	C	D	E	Y	G	H	Z	J	K	W	M	N	O
Position of the amino acid	39	40	41	42	43	44	45	46	47	48	49	50	51	52	53

Table 2

Animal	Number of Amino Acid Positions Different from Horse
Whale	
Turtle	
Rabbit	
Donkey	
Snake	

Classic **Lab** 16, **How do species compare?** continued

LAB 16   63

Data and Observations

1. In the space below, construct a branching tree using this information. Your tree should include horses, donkeys, snakes, whales, turtles, and rabbits. Your tree should show one way that these species could have evolved from a common ancestor.

Analyze and Conclude

1. Which species are closest to a horse as indicated by the particular sequence of amino acids in **Table 1**?

   _____

   _____

2. Which species are more distantly related to a horse as indicated by the sequence of amino acids in **Table 1**?

   _____

   _____

   _____

   _____

3. Based on the evidence in **Table 2**, how would you describe species with similar amino acid sequences in terms of their shape and structure?

_____

_____

_____

4. Based on **Table 2**, make a general statement about whether any of the species other than horses are more or less similar to each other.

_____

_____

5. Based on the comparison you've made, rabbits and whales differ from horses by the same amount. What does this say about the relatedness of rabbits and whales? What might help you understand how those two animals are related? Explain your answers.

_____

_____

_____

6. If you compared the amino acid sequences in **Table 1** with those of another species, and found a different relationship between the species (for example, that snakes were closer to rabbits than to turtles) what conclusions might you draw?

_____

_____

7. **Error Analysis** What are possible sources of error in your experiment? How could you correct for them if you repeated the experiment?

_____

_____

_____

Inquiry Extensions

1. How would you compare the relatedness of all of the animals cited in **Table 1**? How closely related is each of the animals to all of the other animals? Would they change how you draw a branching tree for this group of species? If the data you compile in comparing all of the species seems contradictory (i.e., one species seems closer to another on one chart, but not on another), how would you resolve the contradiction? Would more data help, and if so, what kind of data?

2. What predictions could you make about amino acid sequences for species not shown in **Table 1**? Research amino acid sequences for cytochrome c for a new group of species. Compare the sequences using the same techniques as in this lab. Do your results support your prediction?

Classic
Lab 17

Could you beat natural selection?

Natural selection uses the principle of survival of the fittest. Fitness is often defined as the suitability of an organism to a given environment. It might be the case, however, that a certain set of features or characteristics that are favorable to an organism in one environment might prove to be unfavorable in a different environment. In some cases it might be true that altering the environment of an organism might decrease its chances of survival. In this lab, you will learn more about natural selection and survival of the fittest by deciding which characteristics are more favorable for survival in a variety of environments.

Objectives

- Locate organisms (represented by chips) in the natural environment of the classroom.
- Make a prediction about survivability of two sets of organisms.
- Simulate predator/prey relationships.
- Complete data tables.
- Graph results.

Materials
Part A
clear plastic chips
plastic chips in three additional colors
graph paper
colored pencils
calculator

Part B
one page of newspaper apartment rentals or stock quotes
sheet of plain paper, the same size as the newspaper
envelope of paper circles representing prey
forceps or pencil with eraser
stopwatch or watch with second hand
calculator

Procedure
Part A. Predator/Prey Relationships

1. Read and complete the lab safety form.
2. There are 100 plastic chips hidden around the room. You will have 3 min to search for them. Gather the ones that you find, and note the locations where they were found.
3. Stop after 3 min and count the number of chips that you found.
4. Work with your classmates to tabulate the total number of chips found by the class.
5. Complete **Table 1** showing the following information: original number of chips, color of chips, number of chips found by you, location found, number found by the rest of the class.
6. Use the graph paper to graph the results of the class with your data. In a bar graph, plot your data in one color and the class data in another.

Classic **Lab** 17, **Could you beat natural selection?** continued

Part B. Camouflage

1. Work with a partner. Decide which partner will be the predator and which will work with the prey (the "prey manager"). You must keep these roles throughout the exercise.

2. The predator hunts at twilight and in the early evening. The prey are the newsprint circles and the plain circles. These two sets of circles live in two different environments—newsprint paper and plain paper. The predator does not prefer one kind of circle over the other and simply feeds on any circles it come across.

3. With your partner, come up with a prediction explaining how the newsprint circle and the plain circle will be consumed or conserved. Write your prediction in the appropriate spot in the *Data and Observations* section.

4. The predator should wait in the hallway until called in by the prey manager.

5. **Figure 1** The prey manager should distribute both sets of circles randomly over the printed sheet of paper. This person needs to ensure that the circles are not piled up on each other and that they are distributed evenly over the paper.

6. When the prey has been distributed, the prey manager should bring in the predator. The predator should look at the paper and, using forceps or a pencil eraser, count how many circles of each type he or she can pick up in the span of 10 s. The prey manager should keep time.

7. After 10 s have passed, the predator should call out the number of newsprint circles and plain circles he or she picked up, and the prey manager should record these numbers in **Table 2**.

8. **Figure 2** The team should prepare for a second feeding. This time the predator should cover his or her eyes while the prey manager places both sets of circles on the plain paper. Again, count the number of circles the predator picks over the span of 10 s.

9. Repeat steps 4–8 two more times. Between each pair of trials, have the predator return to the hallway.

10. When the trials are finished, return the prey to the envelope and return all the supplies to your teacher.

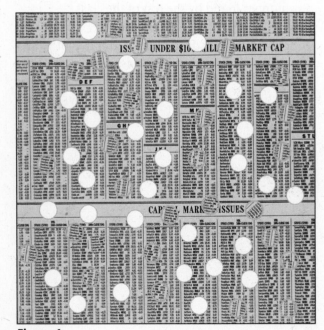

Figure 1

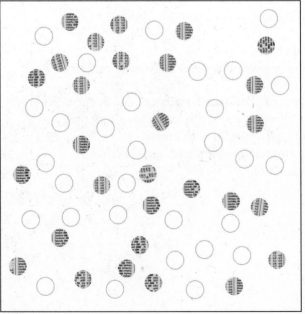

Figure 2

Classic **Lab 17, Could you beat natural selection?** continued

Data and Observations
Table 1

Chips Data					
Chip	Original Number	Number Found by Me	Total Number Found	Number Left	Percentage of Chips Left
Clear					
Red					
Yellow					
Blue					

Prediction for Part B:

_____

_____

Table 2

Circles Data	Plain Background		Newspaper Background	
	Plain Circles	Newsprint Circles	Plain Circles	Newsprint Circles
Total population				
	Number of Plain Circles Consumed	Number of Newsprint Circles Consumed	Number of Plain Circles Consumed	Number of Newsprint Circles Consumed
Trial 1				
Trial 2				
Trial 3				
Team average				
Percentage of circles that died				
Percentage of circles that survived				

Analyze and Conclude

1. Which of the four kinds of chips were most easily found?

_____

2. Chips of which color were most difficult to find?

_____

3. What environmental factors in the room allowed some protection for the chips?

_____

4. Analyze which characteristics were favorable for these organisms and which characteristics made their survival less likely? Explain your answers.

_____

_____

5. For Part B, how do the phenotypes of each species of circle affect the survival of the organisms?

_____

_____

_____

6. **Error Analysis** What are some possible sources of error in your experiment?

_____

_____

_____

7. After completing Part B of the laboratory, what can you conclude about the role of an organism's surroundings on its survival? Was this demonstrated by your experiences in Part A?

_____

_____

_____

Inquiry Extensions

1. This lab has shown a relationship between coloration and natural selection. What organisms use camouflage to increase their chances of survival? Create an electronic slide show to demonstrate and explain how these adaptations benefit the organism.

2. What other phenotypes (outward appearances) or physical characteristics contribute to an organism's survival? Given what you know about trends in environmental conditions, write a description of an adaptive characteristic that would benefit, during the next 100 years, an existing species that you encounter near your home.

Design Your Own
Lab 18

Does this animal walk on four legs or two?

Primates, early and present-day, have several characteristics in common. They share the ability to see in color and grasp with their five-digit hands, and the tendency to care for their young. Primates are divided into two groups, strepsirrhines/haplorhines anthropoids, a subgroup of the haplorhines Humans and apes are part of the anthropoid group. Evidence suggests that early hominoid anthropoids started walking on two legs about 4 million years ago.

In order for bipedalism to occur, several structural changes had to evolve. One of these is that the center of gravity of the hominid organism had to change in order for the body to be balanced in an upright position. A change in the position/shape of the pelvic bones to provide placement for upright walking muscles and support for inner organs was also needed. Finally, the position of the head on the spinal cord is changed. This allows the legs of biped organisms to be directly under the body to support the entire weight of the organism. In this laboratory exercise, you will design your own procedure for comparing the differences between animals that are bipedal and those that walk on four legs.

Problem
Paleoanthropologists need to determine if bone fragments come from an ape or a hominid.

Objectives
- Devise a plan to compare the limbs and pelvis of a gorilla, australopithecine, and present-day human.
- Make predictions about the differences between structures of bipedal and quadrupedal animals.
- Extrapolate the information gathered to a discussion on natural selection.
- Determine whether the australopithecines were habitual upright walkers (bipeds).

Safety Precautions

Possible Materials
calculator
human skeleton
diagram of pelvis and femurs of three animals
diagrams of gorilla, australopithecine, and human
scale drawing of human and gorilla on card stock
glue
scissors
ruler

Hypothesis
Study the diagrams of the human and the gorilla. Write several hypotheses about the bony structures, vertical lines of force, wear patterns, and center of gravity of each organism and how they are designed to best function in that organism.

_____

_____

_____

_____

Plan the Experiment

1. Read and complete the lab safety form.

2. Choose which parts of the body you will measure. Be sure to measure the body parts mentioned in your hypothesis.

3. Decide on your procedure for comparing the skeletons of humans and their early ancestors. In the space provided below, write your procedure for collecting this information. Include any materials you will use.

4. Devise a procedure for determining the center of gravity in each specimen.

5. Decide how you will record your data, and design a data table to hold this information.

Check the Plan

1. Make sure your teacher has approved your experimental plan before you proceed.

Record the Plan

Use the space below to write down your experimental procedure.

Design Your Own **Lab** 18, **Does this animal walk on four legs or two?** continued

Data and Observations

1. In the space below, create a data table of your findings.

Analyze and Conclude

1. Think about the gorilla, australopithecine, and human skeletons that you compared. An animal with a center of gravity above and in front of the pelvis has a tendency to fall over if standing on two legs. Which animal might have this problem? What feature helps it compensate for this?

_____

_____

_____

_____

2. How is body weight distributed in a human? In a gorilla?

_____

_____

3. Based on what you have seen, do you anticipate that the australopithecine walked on two legs or on four? What information led to your conclusion?

_____

_____

_____

4. How might natural selection have caused the change from quadrupedalism to bipedalism?

_____

_____

_____

5. Error Analysis What are some possible sources of error in your measurements?

_____

_____

6. Exchange your plan and data with another group in your class for peer review. What do their data show about the comparisons between these three animals?

_____

_____

_____

Write and Discuss

Write a short paragraph describing your findings and indicating whether or not they support your hypothesis. Discuss any questions your results might have raised.

_____

_____

_____

_____

Inquiry Extensions

1. Scientists who are studying the reasons why human ancestors began walking upright are focusing on some key factors. Research and report on the hypotheses that are currently being explored. What evidence is there to support these hypotheses?

2. Create a family tree tracing the distant relatives of humans and how they have changed over time. Include a diagram of each relative, if possible, and a short paragraph about their distinguishing characteristics. Present your information to your class in the form of a poster or time line.

Classic
Lab 19

What is a taxonomic key?

Classification is a way of separating a large group of closely related organisms into smaller subgroups. The scientific names of organisms are based on the classification systems of living organisms. To identify an organism, a scientist might use a key. A key is a listing of characteristics, such as structure or behavior, organized so that an organism can be identified.

In this lab you will create a taxonomic key for the order Artiodactyla. Artiodactyls are mammals whose feet have an even number of toes, also known as paraxonic feet. Artiodactyls are primarily herbivores. This is a large and diverse group of mammals. There are approximately 220 living species of artiodactyls. Most live in open plains or savannas, but others live in forests, and some are semi-aquatic. Some of the fastest-running mammals are found in this order, but there are some that are slow and clumsy as well.

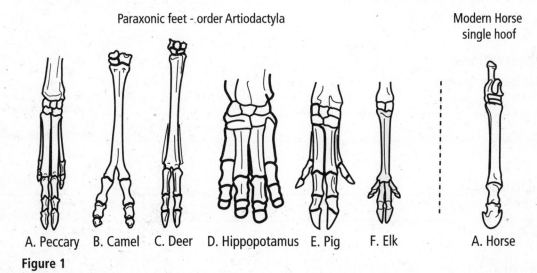

Paraxonic feet - order Artiodactyla   Modern Horse
single hoof

A. Peccary   B. Camel   C. Deer   D. Hippopotamus   E. Pig   F. Elk   A. Horse

Figure 1

Objectives
- Use a key to identify common denominations of money.
- Examine the method used to make a key.
- Construct a key to identify a group of organisms.

Materials
set of coins (penny, nickel, dime, quarter) and bills ($1 and $5)
pen or pencil
reference material
animal envelope
index cards

Safety Precautions

Procedure
Part A. A Simple Taxonomic Key
1. Read and complete the lab safety form.
2. In the *Data and Observations* section you will find an unfinished taxonomic key for the money you have in front of you.
3. Fill in the missing information in the key, and pay attention to the way that the taxonomic key is set up and what sort of information is in it. You will be making your own taxonomic key in the next section.
4. Return all money to your teacher.
5. Wash your hands with soap and water after completing this part of the activity.

Classic **Lab** 19, **What is a taxonomic key?** continued

Part B. Making Your Own Taxonomic Key

1. Retrieve an animal envelope from your teacher.
2. Inside are pictures of ten animals belonging to order Artiodactyla. These animals are the even-toed ungulates. Ungulates are mammals with hooves, like horses. To see how even-toed ungulates differ from odd-toed ungulates like horses, see **Figure 1**.
3. Work with a partner to design a taxonomic key that will list the characteristics of these animals in a way that the organism can be identified.
4. Write down some characteristics of each animal. Pay attention to those characteristics that can distinguish one animal from another. Start with the most general characteristics and progress to increasingly more specific characteristics. Avoid using descriptors such as "large" or "small," if possible.

5. Write the characteristics you see on index cards. This will make them easier to manipulate and organize.
6. Determine which characteristic gives you the smallest number of subgroups. This is a good starting point for the key.
7. Determine how to break each subgroup into smaller subgroups, using pairs of characteristics (look back at the key for money). Keep working until you have separated all of your animals into their own groups. Start the choices in a pair with the same word, if possible. Start each couplet with different words, if possible.
8. Keep in mind that not everyone's key will be the same.

Data and Observations

1. Taxonomic Key for Money

1A. Metal ....Go to statement 2

1B. Paper...Go to statement 5

2A. Brown (copper).. _____

2B. Silver ....Go to statement 3

3A. Smooth edge .... _____

3B. Ridges around the edge...Go to statement 4

4A. Torch on back .... _____

4B. Eagle on back .... _____

5A. Number 1 in the corners .... _____

5B. Number 5 in the corners .... _____

Classic **Lab** 19, **What is a taxonomic key?** continued

2. **Taxonomic Key for Order Artiodactyla**

1A. _____ .... _____
1B. _____ .... _____

2A. _____ .... _____
2B. _____ .... _____

3A. _____ .... _____
3B. _____ .... _____

4A. _____ .... _____
4B. _____ .... _____

5A. _____ .... _____
5B. _____ .... _____

6A. _____ .... _____
6B. _____ .... _____

7A. _____ .... _____
7B. _____ .... _____

8A. _____ .... _____
8B. _____ .... _____

9A. _____ .... _____
9B. _____ .... _____

10A. _____ .... _____
10B. _____ .... _____

Analyze and Conclude

1. What is a classification key, and how is it used?

_____

_____

2. List four different characteristics you used in your taxonomic key for the order Artiodactyla. Why did you choose these characteristics?

_____

_____

_____

_____

3. Which main characteristic could be used to distinguish a pronghorn sheep from a bushbuck?

_____

_____

4. Which main characteristic could be used to distinguish a mountain goat from a sheep?

_____

5. Exchange your taxonomic key with that of another pair of students. Work through it to identify the animals. Is the key correct? How does your classmates' key differ from your own?

_____

_____

6. **Error Analysis** What are some possible sources of error in your taxonomic key? What information would have made it easier to overcome these issues?

_____

_____

_____

Inquiry Extensions

1. Choose another order, such as order Primate, and create a taxonomic key for several organisms in that order. Present your key to the class.

2. Walk around your neighborhood or school yard. Choose a category of items about which you can make a taxonomic key. This could be the types of trees or rocks, birds that live in or migrate through the area, or even vehicles in your neighborhood. Draw a detailed poster of the taxonomic key.

Design Your Own
Lab 20

Can you filter out cholera?

In areas of southern Asia, such as Bangladesh, cholera is a common, and often deadly, disease. Copepods (miniature aquatic crustaceans) found in river water can carry a large number of *Vibrio cholera* bacteria in and on their bodies. When people drink untreated river water, the bacteria can produce a toxin that causes the small intestine to secrete massive amounts of fluid rich in salts and minerals, leading to dangerous bouts of diarrhea and dehydration. This disease is called cholera.

The river water can be sterilized by boiling, but this is not generally done because wood for fuel is scarce in Bangladesh and in many other developing countries. Tests have shown that filtering out the copepods can also remove much of the bacteria.

Recently, scientists and residents of Bangladesh have found that simple filters made of the cloth from women's saris can reduce the number of cholera cases by up to 50 percent. A sari is a traditional woman's garment made from a single rectangular piece of cloth measuring 5-6 m in length. It can be made from cotton, silk, or synthetic materials.

In this lab, you will use different types of fabrics to create your own water filter that could be used to clean the water in this area.

Problem
Test filters made of common cloth to see which will remove at least 25 percent of the copepods from river water.

Objectives
- Form a hypothesis about what type of filter would be the best to filter water containing copepods.
- Design a filter.
- Compare the number of copepods in a water sample before filtering to the number of copepods after filtering.

Possible Materials
simulated river-water sample
funnel
clean droppers (2)
large beaker
small beaker
graduated cylinder
microscope
slides with cell counter or slides with grids
cover slips
metric ruler
squares of cloth

Safety Precautions

Hypothesis
Use what you know about copepods, bacteria, and filters to write a hypothesis that will explain how to optimize copepod removal from a river sample.

_____

_____

_____

_____

Plan the Experiment

1. Read and complete the lab safety form.
2. Choose the material, or materials, you will use to make your filter.
3. Decide on a procedure for counting the copepods in the water before it is passed through the filter and after the water passes through the filter. Write your procedure for counting the copepods and setting up the experiment in the space provided below.
4. Identify the independent variable, dependent variable, constants, and control group.
5. Decide how you will record your data and when you will record it. Design a data table to hold your data and observations.

Check the Plan

1. Make sure the teacher has approved your experimental design before you begin. Make sure your personal safety equipment, including goggles, apron, and gloves, is in place before starting the experiment.
2. Be sure that a control group is included in your experiment and that the experimental group varies in only one way.
3. When you have completed the experiment, dispose of the simulated river water as directed by your teacher.
4. Wash your hands with soap and water.

Record the Plan

In the space below, write out your plan for testing the materials as filters. Sketch how you plan to set up the filters to allow the water to pass through.

Design Your Own **Lab** 20, **Can you filter out cholera?** continued

Data and Observations

1. Use the space below to create a data table of your findings.

Analyze and Conclude

1. Briefly explain your reasoning behind the construction of your filter.

   _____

   _____

   _____

2. How well did your setup remove the copepods from the water? What was the
   percentage of difference in the number of copepods at the beginning of the
   experiment and the number at the end?

   _____

   _____

   _____

3. What could be other benefits to using this type of filter? Explain.

   _____

   _____

   _____

4. Describe the control in your experiment. What did the control show?

_____

_____

_____

5. Error Analysis What were some possible sources of error in your experiment?

_____

_____

_____

_____

6. Exchange your procedure and data with another group in your class for peer review. What do their data indicate?

_____

_____

_____

_____

_____

Write and Discuss

Write a short paragraph describing your findings, and indicate whether or not they support your hypothesis.

_____

_____

_____

_____

_____

Inquiry Extensions

1. Suppose your local water supply might be contaminated due to flooding. Design an experiment to determine whether the water is contaminated. Use the knowledge you have gained about filters. Show your design to the class.

2. The sari filters work well, but the procedure for making and using the filter needs to be distributed to everyone. Plan a public information campaign aimed at lowering cholera rates. Choose the best means of communicating, and decide what your main message will be.

Design Your Own
Lab 21

Do protists have good table manners?

Protists are eukaryotes that are not animals, plants, or fungi. Protists vary in structure and function more than any other group of organisms. Most protists are unicellular, although some, such as seaweed, are multicellular. Protists have organelles and a nucleus with a nuclear envelope.

Generally, protists are characterized by the types of food they consume and how they obtain it. Animal-like protists are heterotrophs that ingest food that they come across in the environment. Funguslike protozoans are also heterotrophs, but they eat mainly decaying organic matter. Plantlike protozoans are autotrophs and make their own food.

The manner in which protozoans eat also makes them unique. Some absorb food through their cell membranes. Others, like amoebas, surround food and engulf it. Others have openings called mouth pores into which they sweep food. In this lab, you will observe the eating techniques of paramecia—a type of protist.

Objectives
- Form a hypothesis about how environmental factors impact the eating habits of paramecia.
- Observe paramecia eating under a microscope.
- Identify a variable to test.
- Introduce an environmental variable, and record any changes in the paramecia's eating habits.

Possible Materials
microscope
slides and cover slips
eye droppers (2)
sample of pond water
yeast mixture
data table
table lamp
methyl cellulose or 3% gelatin solution
ice
plastic gloves
warm water (26°C to 30°C)
cold water (14°C to 18°C)
caffeine solution (1g/L)
sugar

Safety Precautions

WARNING: *Use caution when handling slides— broken slides can cut skin. Be sure to keep lamps away from water sources to avoid shock.*

Hypothesis
Use what you know about the movements and behaviors of paramecia to write a hypothesis indicating the impact of an environmental factor on the way paramecia consume their food.

_____

_____

_____

_____

Plan the Experiment

1. Read and complete the lab safety form.

2. Obtain a sample of pond water from your teacher.

3. Decide how you will identify specific paramecia to study and how much food you should give them. If the paramecia are moving too quickly for you to observe, slow them down by adding methyl cellulose or 3-percent gelatin solution to the water.

4. Decide which environmental factor you will change while you observe their eating habits.

5. Decide how you will observe the paramecia eating and how you will record your observations.

6. Discuss your plan with your classmates and teacher.

7. Try drawing what you see under the microscope. Design a data table to hold your observations.

8. Decide if there are other variables that could control the eating habits of paramecia. What roles do temperature, sunlight, the amount of food available, or competition from other paramecia play in determining how the organism eats? Choose a variable you want to test.

9. Identify the independent variable, dependent variable, constants, and control group for your new test, and record your data. Share your results with the class.

Check the Plan

1. Be sure that you have included a control group for the second part of the exercise.

2. Wear gloves when handling the culture.

3. Make sure your teacher has approved your procedure for the second part of the lab before you proceed.

4. Observe the behavior of a paramecium as it eats.

5. When you have completed the experiment, dispose of the pond water as directed by your teacher.

Record the Plan

In the space below, write your experimental procedure.

Design Your Own **Lab 21, Do protists have good table manners?** continued

Data and Observations

1. In the space below, make drawings to show what you saw through the microscope as the paramecia ate the yeast mixture. Include a labeled diagram of a paramecium in your response.

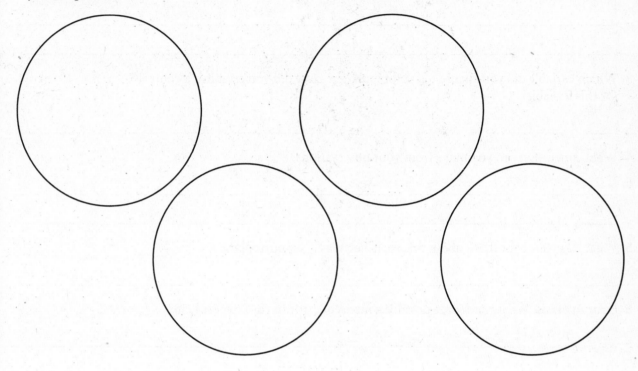

2. Use the space below to list the different behaviors that the paramecia of the test group exhibited as they ate.

3. Use the space below to explain how the behavior of the eating paramecia changed when one variable was changed.

Analyze and Conclude

1. How do paramecia eat? Describe your observations.

_____

2. Based on your observations, what is the role of the cilia in the consumption of food?

_____

_____

3. Which variable did you choose to explore further? How did you establish a control for this variable?

_____

4. What conclusion can you make from your observations?

_____

_____

5. What was your hypothesis about this variable? Was it supported?

_____

6. **Error Analysis** What were some possible sources of error in your experiment?

_____

_____

7. Share your conclusions and observations with your classmates. What inferences can be made about the optimum conditions for paramecia feeding?

_____

Write and Discuss

Write a short paragraph describing your findings and indicating whether or not they support your original hypothesis. Discuss any questions your results might have raised.

_____

_____

_____

Inquiry Extensions

1. How do other protists ingest food? If there are any differences in how food is ingested, how do you explain them? Observe other protists found in your pond water, and compare their methods of feeding with that of the paramecia.

2. Which do you eat more rapidly—french fries or carrots? Do paramecia change their eating habits with different food sources? Research different potential food sources for paramecia. Conduct the experiment again to see if you can observe changes in activity based on the presence of different food sources.

Classic Lab 22

What are mushroom spores?

Mushrooms, such as those that are found in your yard or in the woods, are fungi. Fungi exist in a variety of sizes, ranging from a single cell of yeast to large, multi-cellular mushrooms. Fungi need moist, warm places to grow. All fungi, including mushrooms, are eukaryotes and heterotrophs, and they use spores to reproduce. In this lab, you will examine the characteristics of some common mushrooms and learn how they are spread.

Objectives

- Identify the parts of a variety of supermarket mushrooms.
- Learn about the spores of a mushroom by creating and examining a spore print.
- Determine how spores are spread by making a model from a balloon and cotton balls.

Materials

magnifying lens
mushrooms of different varieties purchased from local grocery store
dissecting probe
paper towels
white paper
large plastic container with lid
round balloon
cotton balls
tape
ruler or stiff stick
modeling clay
pin

Safety Precautions

WARNING: *Do not eat any of the mushrooms you are using in this laboratory.*

Procedure

Part A. Identify the parts of a mushroom.

1. Read and complete the lab safety form.
2. Obtain three mushroom samples from your teacher. Remember, do not eat anything given to you in a laboratory setting. Also, do not eat mushrooms you might find in the wild. Many could be poisonous.

Figure 1

3. **Figure 1** Identify the different parts of each mushroom: the gills, cap, and stalk.
4. Twist off the cap of each mushroom and break open the stalks from end to end.
5. Draw detailed diagrams of each mushroom and label the individual parts of each one. Be sure to include a description of the threadlike structures within the stalk.
6. Dispose of the mushroom pieces as instructed by your teacher and wash your hands with soap and water.

Part B. Make A Spore Print

1. Obtain some mushrooms from your teacher. Gently twist the caps off of the mushrooms.
2. Remember: Do not eat the mushrooms provided in this lab.
3. Cut a piece of white paper to fit and place it inside the bottom of the plastic container.
4. **Figure 2** Place the mushroom cap, gill side down, on the paper and put the cover on. Place the container in an area of the classroom where it will remain undisturbed.

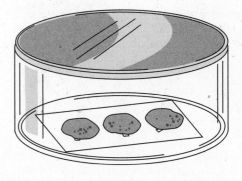

Figure 2

5. Wash your hands with soap and water.

6. After at least two days, carefully remove the top from the container and pick up the mushroom cap. You will find a spore print on the white paper.

7. Examine the spore print with your magnifying lens or a binocular scope if one is available. Draw a detailed diagram of your spore print, describing the relationship between the spores and the structures in the mushroom cap

8. Dispose of the print and mushroom cap as directed by your teacher. Wash your hands with soap and water.

Part C. Spore Dispersal

1. Now that you have seen what the spores of a mushroom look like and where they are stored, make a model of how spores are released into the air.

2. Assemble the materials that are needed for this model. These include the cotton balls, balloon, tape, ruler, clay, and pin.

3. Break a cotton ball into small pieces. Roll them into little balls.

4. Place each small cotton ball into the balloon. Continue until the balloon is about ¾ full.

5. Inflate the balloon, being careful not to inhale any of the cotton ball pieces. Tie off the end of the balloon with a knot.

6. Tape the knotted end of the balloon to the stick or ruler. Stand the stick up in the clay.

7. Make a drawing of your model. Be sure to label the parts of your model with the actual part of the mushroom that they represent.

8. Choose one group member to pop the balloon. The rest of the group members should stand back about one meter from the setup as that person pops the balloon with the pin. (Use caution: Pins are sharp and they can puncture skin.)

9. Observe what happens when the balloon pops.

Data and Observations

1. Use the space below to draw your observations from Part A of the investigation.

2. Use this space to draw your mushroom spore print.

Classic **Lab 22, What are mushroom spores?** continued

3. Use this space to draw and label your experimental design for spore dispersal and the results.

Analyze and Conclude

1. What are the threadlike structures inside the stalk of the mushroom? Of what are they made? Did every mushroom you examined contain them? Explain.

_____

_____

2. What function do you think these structures serve underground for the mushroom?

_____

_____

_____

3. Look again at the diagram you drew of your spore print. Based on what you saw in the print, how many spores do you think a mushroom could produce? Where do you think mushroom spores might be most likely to grow into new mushrooms? What do you think happens to the spores that do not grow into mushrooms?

_____

_____

_____

_____

4. Based on what you saw in your spore print, and in your model of spore dispersal, why do you think mushrooms are found just about anywhere?

_____

_____

_____

5. How did building a model help you understand the dispersal of mushroom spores better?

_____

_____

_____

6. **Error Analysis** What are possible sources of error in your experiment?

_____

_____

_____

7. How might the addition of wind, in the form of a fan, affect the results of your spore dispersal model?

_____

_____

_____

_____

Inquiry Extensions

1. There are more than 3000 types of mushrooms in North America. Research one poisonous variety. Be sure to include a diagram of the major structures as well as hints as to how to identify it.

2. Mushrooms have a wide array of methods and mechanics for releasing spores. What are some of these methods? Besides wind, what other factors might aid in spore dispersal? What other kinds of dispersal methods do mushrooms use?

How do ferns, mosses, and conifers reproduce?

Have you seen recent films or TV shows featuring computer-generated dinosaurs? The most accurate depictions show dinosaurs tramping through fern and pine forests and strolling over moss-covered ground. These types of plants were abundant during the Mesozoic era (248 to 65 million years ago).

The seeds, spores, and sperm of different types of plants vary widely in structure and function. In this laboratory experiment, you will examine the actual spores and seeds of a variety of plants and then examine diagrams of the life cycles of these plants. This information will be used to compare the life cycles of ferns, mosses, and conifers.

Ferns are seedless vascular plants. As you examine the fern fronds, you will notice small brown or black dots on the underside. These are spore capsules (sori). Mosses are nonvascular plants. They rely on osmosis and diffusion to transport water and nutrients from one part of the plant to another. Conifers are gymnosperms that produce both pollen and seeds.

Objectives

- Examine samples of ferns, mosses, and conifers.
- Compare characteristics of seeds, spores, and pollen in each.
- Infer how these characteristics have made survival of each plant possible.

Materials

fern fronds
moss sample
pine cones (male and female)
diagrams of the life cycles of ferns, mosses, and conifers
scalpel
forceps
paper towels
dropper
water
petri dish
magnifying lens
colored pencils

Safety Precautions

WARNING: *Use extreme care when handling the scalpel—it is very sharp and can easily cut or puncture the skin.*

Procedure

1. Read and complete the lab safety form.
2. Examine the diagram of the life cycle of each type of plant.
3. Obtain samples of a fern, moss, and pine cones. Examine each one closely, using the magnifying lens or stereomicroscopes if necessary. Make a detailed drawing of each in the *Data and Observations* section of this lab.
4. As you make your drawings, label the following as you see them: seeds, spores, sperm, ovaries, and pollen.
5. After completing your observations and drawings, identify one spore capsule on a fern frond and one seed from the pine cone. Isolate these and place each on a piece of paper towel. If your moss sample has well-developed sporophytes, you can examine spores of these as well.
6. Working carefully, use the scalpel to open either the seed, sporophyte, or spore capsule. Draw what you see. Notice the stage of development of each and make note of any other characteristics you see. Make a detailed drawing of each.
7. Clean up your lab station as directed by your teacher. Wash your hands with soap and water after handling the specimens.
8. Refer to your drawings when answering the questions that follow.

Data and Observations

1. In the space below, make detailed drawings of the fern frond, moss sample, and pine cone.

Fern Frond

Moss

Pine Cone

Classic **Lab 23, How do ferns, mosses, and conifers reproduce?** continued

2. In the space below, make detailed and labeled drawings of the spore capsule, moss sporophytes, and the conifer seed.

Spore Capsule

Moss Sporophytes

Conifer Seed

Analyze and Conclude

1. What features did you see on the fern frond?

_____

_____

2. What did you notice about the moss sample?

_____

_____

3. What features did you see in the pine cone?

_____

_____

4. What do the differences in the structures of these three plants tell you about the life cycle of each one?

_____

_____

5. How is a fern capsule different from a conifer seed? How are they similar? How does a moss sporophyte compare to the other two?

_____

_____

6. **Error Analysis** What were some possible sources of error in your analysis?

_____

_____

_____

7. What conclusions can you draw about the survival needs of each type of plant based on your observations?

_____

_____

_____

Inquiry Extensions

1. What structures in a moss plant make it possible for it to live on land? Explore a sample of moss in more depth, and design an experiment to examine the role moisture plays in the survival of mosses.

2. What role do these three kinds of plants play on Earth? What is their value in an ecosystem? What is their value to human beings?

Design Your Own
Lab 24

Do plants sweat?

Imagine that a catastrophic change in weather patterns caused a shift in air currents that has virtually eliminated rainfall over the rain forests of Costa Rica. Rainfall that was normally 37 cm per month has dropped to 5-10 cm per month, and coastal winds have increased. Now instead of a steady rain, plants depend solely on water that they receive from rivers and groundwater.

Water moves within the roots, stems, and leaves of a plant. Transpiration is the loss of water through leaves due to evaporation. Transpiration makes use of cohesive and adhesive forces on the water within the xylem to bring water from the ground and raise it through the trunk, branches, stems, and leaves. In this way, the plant can deliver water and dissolved nutrients to all of its parts and cool itself.

In one part of the rain forest, multiple streams and rivulets supply adequate water to the roots of trees and ground-hugging plants. It is also shaded for a good portion of the day and protected from winds by cliffs on three sides. Is there any chance that this area might begin to recapture the humid atmosphere of the old rain forest? You have been assigned to research how effectively the rain forest plants are adjusting, or not adjusting, to the new conditions.

Problem
Plants that used to be in a humid environment are now in a dry environment.

Objectives
- Form a hypothesis about how transpiration is affected by a change in an environmental condition.
- Design an experiment to test the impact of this environmental condition.

Safety Precautions

WARNING: *Keep fans away from water, and plug them into a GFI-protected circuit. Handle sharp scissors with care. Use only fresh plants—plants placed in water over several days will have bacterial or mold growth.*

Possible Materials
electric fan
food coloring
live plant
pipette
scissors
small beaker
small plastic bags (not the zippered kind)
ties
petroleum jelly
water

Hypothesis
Use what you know about transpiration to write a hypothesis that could explain which environmental factors affect the rate of transpiration in a plant.

_____

_____

_____

Plan the Experiment

1. Read and complete the lab safety form.
2. Choose which environmental factor you will investigate. Transpiration rates can be influenced by humidity, the amount of light, wind, and other factors.
3. Decide on a procedure to use to test the impact that the environmental factor has on transpiration. You might have to observe your design over the course of several days. Record your procedure below. Include the materials you will use.
4. Identify the independent variable, dependent variable, constants, and control.
5. Sketch the experimental setup.
6. Design a data table to record environmental factors and information on the amount of transpiration. Include a time line describing the duration of the experiment and individual checkpoints.
7. The structure of a plant stem has a big impact on transpiration within the plant. Include a labeled diagram of the structure of the plant stems you are studying, and use this diagram to indicate the significance of the movement of water within the plant.

Check the Plan

1. Be sure that a control group is included in your experiment and that the experimental group differs in only one way.
2. Make sure your teacher has approved your experimental plan before you proceed.
3. Observe your experimental design for evidence of transpiration.
4. When you have completed the experiment, dispose of the plant cutting as directed by your teacher.

Record the Plan

In the space below, write your experimental procedure and make drawings of your experimental design. Be sure to indicate your control, the variables, and the constants in your design.

Design Your Own **Lab** 24, **Do plants sweat?** continued

Data and Observations

1. Use the space below to create a data table of your findings, including the location of the plants, the variables measured, and the transpiration observed.

Analyze and Conclude

1. Which environmental factor did you choose to examine in this experiment? Why did you choose this one?

_____

_____

_____

2. Based on your observations, how did your chosen environmental factor affect the rate of transpiration?

_____

_____

_____

3. What other factors do you think might affect the rate of transpiration in a plant? How might you test for those?

_____

_____

_____

4. Describe the control in your experiment. What did the control show?

_____

_____

_____

5. Error Analysis What were some possible sources of error in your experiment?

_____

_____

_____

_____

6. Exchange your procedure and data with another group in your class for peer review. What does their data show about transpiration in plants?

_____

_____

_____

_____

Write and Discuss

Write a short paragraph describing your findings and indicating whether or not they support your hypothesis. Did the environmental factor you studied affect transpiration? Discuss any questions your results might have raised.

_____

_____

_____

_____

_____

_____

Inquiry Extensions

1. Examine plants in your yard or in a local park. What environmental factors exist in these locations that could affect the rate of transpiration for these plants? Make a map of the area which shows the factors that could affect it.

2. In addition to water, humans secrete some chemicals that they have metabolized in their sweat. How would you determine if plants have the same ability? Conduct the experiment again to see if you can detect chemicals released during transpiration. Present your findings to the class.

Classic Lab 25

How does a flower grow?

Flowers come in many shapes and sizes, but they all perform the same function—reproduction. A flower is the reproductive structure of an angiosperm. Sepals, petals, stamens, and pistils are easily identifiable structures within a flower. In this lab, you will examine each of these carefully and determine how they relate to the overall function of the flower.

Objectives

- Dissect flowers to examine female and male parts.
- Measure and describe characteristics of flowers.
- Draw and label flower diagrams.
- Draw conclusions about reproduction in plants.

Materials

paper towels
dropper
magnifying lens
flower identification book (field guide)
microscope
slide
large flowers (2)
cover slip
scalpel
cellophane tape
water
metric ruler
lens paper
colored pencils

Safety Precautions

WARNING: *Be careful when using the scalpel. It is extremely sharp and can cut or pierce the skin.*

Procedure

Part A. The Visible Parts of the Flower

1. Read and complete the lab safety form.
2. Cover your work space with a few paper towels. Obtain a flower from your teacher (Flower 1).
3. Carefully examine the flower. Observe, measure, and record as many characteristics about your flower as possible. These could include the color of the petals, number of petals, distinguishing marks, scent, and length of the petals and flowers. Record this information in **Table 1**.
4. Create a sketch of your flower in the space provided on the next page, label the parts, and list your observations and measurements next to it. Take special care to notice the sepals and the relationship they have to the rest of the flower. Be sure to include these in your diagram.
5. Carefully pull away the sepals to expose the bases of the petals underneath. (If you have difficulty removing the sepals, you can use the scalpel to cut them away as long as you are careful not to damage the underlying structures.) Make special note on your diagram of the number of petals and any differences between different petals.

Part B. The Male Part of the Flower

1. Carefully pull off all the petals of the flower. This will expose the male parts of the flower.
2. Locate the stamens. Draw a diagram of what you see. Measure the length of each stamen, and indicate this on your diagram. Also include information on the number and shape of the stamens.
3. Use the scalpel and carefully cut the stamens away from the rest of the flower. Place these on your paper towel.
4. Obtain a clean microscope slide and cover slip. Gently tap some of the pollen grains from the anther onto the slide. Make a wet mount of the pollen.
5. Observe the pollen you collected at low power and high power under your microscope. Draw what you see under the microscope in the space provided in the next section. Be sure to indicate the magnification you are using.

Part C. The Female Part of the Flower

1. Carefully cut the pistil away from the rest of the flower using the scalpel. Write down as many measurements and descriptions of the pistil as possible. Include a diagram of the pistil with your explanation.

2. Check for other properties of the pistil. Check to see if the top of the pistil will pick up a piece of lens paper. Record your observations near your diagram.

3. Lay the pistil on the paper towel, and use the scalpel to carefully cut it in half. Draw a diagram of what you see, including the number of compartments and ovules.

Part D. Flower Comparisons

1. Obtain another flower from your teacher (Flower 2). Repeat steps A, B, and C, and compare the new flower with the original flower you examined. Draw and label diagrams of what you see.

2. Use the field guide to identify each flower.

3. Answer the questions in the *Analyze and Conclude* section.

Data and Observations
Table 1

Flower Data			
Part of the Flower	**Flower 1**	**Flower 2**	**Description**
Petals			
Sepals			
Stamens			
Pistil compartments			
Ovules			

Visible Parts of Flower 1
Male Parts

Female Parts

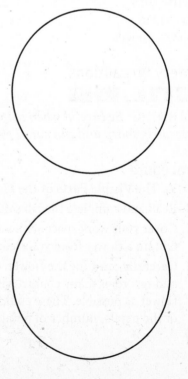

Classic **Lab 25, How does a flower grow?** continued

Visible Parts of Flower 2
Male Parts

Female Parts

Analyze and Conclude

1. How do the sepals relate to the rest of the flower? What did you notice about the sepals of each of your flowers?

_____

_____

_____

2. Were all the petals on your flowers the same? Explain.

_____

_____

_____

3. Based on your observations, how would you describe the arrangement of the sepals, petals, stamens, and pistils in your flowers?

_____

_____

_____

_____

_____

4. What was the relationship between the number of ovules, eggs, seeds, and ovaries in your flowers?

_____

_____

_____

5. What is the function of the flowers you observed? How are the sepals, petals, stamens, and pistils arranged to serve this function?

_____

_____

_____

6. You measured the height of the pistil and the stamens in each flower. Why is the pistil taller? How could this help you explain how this flower is pollinated?

_____

_____

_____

7. Error Analysis What are some possible sources of error in this experiment?

_____

_____

_____

_____

8. Were your flowers monocots or dicots? How do you know?

_____

_____

_____

Inquiry Extensions

1. How can the examination of a flower tell you anything about the plant from which it came? Use your observations from this lab to write a paragraph explaining the connection between flowers and plants.

2. Examine fruit provided by your teacher. Isolate the seeds and, if possible, dissect them to see the cotyledons. Using your knowledge of the growth habit of the parent plant and what you have seen of the fruit and seeds, make a prediction about the characteristics of the fruit's flower, methods of pollination, and seed dispersal. Follow up with research in the library. What role does the fruit perform? In its natural habitat, is the method of seed dispersal efficient and effective?

Classic
Lab 26

Is that symmetrical?

Some animals have bilateral symmetry. This means that a line can be drawn through the animal's body that divides the organism into two halves that are mirror images of each other, as shown in **Figure 1**. Nonliving objects such as spoons and eye-glasses have bilateral symmetry. Animals with radial symmetry have many lines of symmetry that pass through a central point, as shown in **Figure 2**. Bicycle wheels have radial symmetry.

The bodies of complex animals all have either bilateral or radial symmetry. In this lab, you will explore the symmetry found in some animals, create models, and investigate the symmetry of human faces.

Figure 1

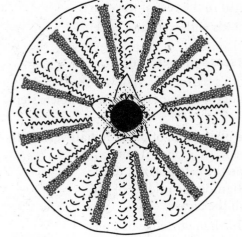

Figure 2

Objectives

- Identify lines of symmetry in animal samples.
- Model symmetry in the human face.
- Infer relationships between body structure and survival.

Safety Precautions

WARNING: *Handle the animal samples with care. Wash hands with soap and water after handling the specimens; preservatives can be toxic. Use caution when cutting the clay with the plastic knife.*

Materials

pencil
ruler
plain, unlined paper
selection of full facial portraits cut in half lengthwise
tracing paper
glue or rubber cement
cellophane tape
preserved sample or photograph of an animal
supplementary resource materials on each animal sample
modeling clay in two colors
plastic knife
small mirror

Procedure

Part A. Types of Symmetry

1. As you conduct your examination of each animal sample, record your observations in the appropriate cell of **Table 1**.

2. Examine each animal sample your teacher has set out. Identify the animal and which type of symmetry it has. Record this in **Table 1**.

3. In **Table 1**, draw a detailed diagram of the animal, indicating the line or lines of symmetry. Use the small mirror to ensure that your diagram is symmetrical.

4. Use any supplementary resource materials you have to fill in the rest of the chart.

5. Wash hands with soap and water.

Part B. Are faces symmetrical?

1. Work with a partner. Each of you should take one half of the same portrait and complete steps 2-6.

2. Fold the plain piece of paper in half vertically to crease it, then open it and lay it flat. Glue the half-portrait to the paper aligning the middle of the face with the crease.

3. **Figure 3**. Align one edge of a piece of tracing paper along the middle of the portrait and tape it in place so that it can move like the page in a book.

4. Fold the tracing paper over the portrait and trace the outline of the face and its features. Draw carefully and accurately.

5. When you have completed the tracing, fold the tracing paper to the empty side of the paper to complete the face.

6. Compare your completed face with the one made by your partner.

Part C. Make it Symmetrical

1. Work with a partner.

2. Each of you will work with different colored clay.

3. Look back at your diagrams from Part A.

4. Choose one animal to model in clay. Mold the shape for half of the animal.

5. Your partner should model the other half of the animal.

6. Fully assemble the new animal and determine if you have re-created the whole animal.

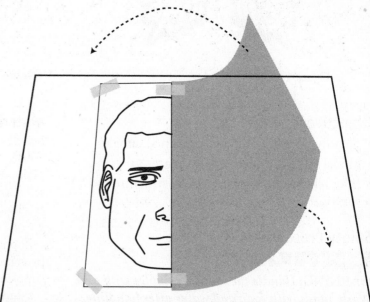

Figure 3

Classic **Lab** 26, **Is that symmetrical?** continued

Data and Observations
Table 1

Symmetry Data				
Animal	**Movement**	**Habitat**	**Symmetry**	**Diagram**

Classic **Lab 26, Is that symmetrical?** continued

Analyze and Conclude

1. Which animals showed bilateral symmetry? Which showed radial symmetry?

_____

_____

2. How do animals with radial symmetry compare with animals with bilateral symmetry?

_____

_____

3. How could bilateral symmetry be advantageous when it comes to escaping from predators?

_____

_____

_____

4. Error Analysis What are some possible sources of error in your activity?

_____

_____

_____

5. Is your face truly symmetrical? Explain.

_____

_____

_____

6. Were you and your partner able to construct an accurate model of the animal? Looking at all the samples in the class, which type of symmetry seems easier to model?

_____

_____

_____

Inquiry Extensions

1. Choose five or six common objects around your home, such as a ladder, a plate, or a tool. Find one example of bilateral symmetry and one of radial symmetry. Write a paragraph describing how the symmetry of each object relates to or supports its function.

2. Observe as many animals as possible in your local area. Keep a list of the animals and later draw diagrams of certain ones and mark their lines of symmetry. Do most animals that you see have bilateral or radial symmetry? Why do you suppose that is?

Design Your Own
Lab 27

Which will the worm choose?

Earthworms, like other animals, have a preference when given the choice between two environmental extremes. Think about your preferences. Do you like to sleep in total darkness, or with a light on? Do you prefer hot temperatures or cold temperatures? In this lab, you will design an experiment that looks at the preferences of earthworms.

Problem
Think about places that you are most likely to see earthworms. Determine if there are certain conditions that earthworms favor over others.

Objectives
- Identify environmental factors that a worm might favor.
- Design a laboratory experiment to determine which condition an earthworm favors.
- Compare the behaviors of two earthworms under a variety of conditions.
- Draw a conclusion about the conditions preferred by these organisms.

Safety Precautions

Possible Materials
stopwatch
earthworms (2)
water
clean spray bottle or dropper
paper towels
shallow pan
cardboard
flashlight
soil
sand
non-mercury thermometer
ruler

Hypothesis
Use what you know about earthworms to write a hypothesis indicating which of a pair of related environmental conditions earthworms prefer.

_____

_____

_____

_____

Design Your Own **Lab** 27, **Which will the worm choose?** continued

Plan the Experiment

1. Read and complete the lab safety form.
2. Decide on a procedure to use to test the preference of the worms. Your teacher will review the proper handling of live animals with you. In addition, refer to **Figure 1**. It demonstrates an acceptable method for keeping earthworms moist. Do not, however, leave the worms in a deep puddle—they will drown.
3. In the space provided below, write your procedure for testing the preference of the animals. Include the materials you will use.
4. Identify the independent variable, dependent variable, control group, and constants.
5. Decide how you will determine which factor the worms prefer. Try watching their behavior over a period of time and watch which environmental factor they tend to move toward. You may want to run your trial twice—or if time allows, several times—to make sure that your observations are accurate.
6. Determine how you will record your data and observations and when you will record it. Create a data table to record your observations of the worms' movements over a period of time.

Check the Plan

1. Make sure your teacher has approved your experimental plan before you proceed.
2. Be sure that a control group is included in your experiment and that the experimental groups vary in one way only.
3. Observe the behavior of your earthworms over a pre-determined period of time.
4. When you have completed the experiment, return the earthworms to their original container and dispose of the other materials as directed by your teacher.
5. Wash your hands with soap and water.

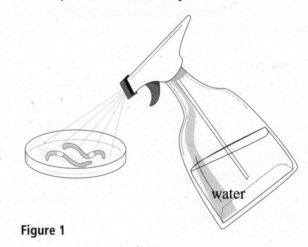

Figure 1

Record the Plan

In the space below, write your experimental procedure and make a sketch of your experimental setup.

Design Your Own **Lab 27, Which will the worm choose?** continued

Data and Observations

1. In the space below, make drawings of the worms at the beginning and the end of the experiment.

Analyze and Conclude

1. Which environmental factor did you choose to investigate? What was it about your knowledge of worm behavior that led you to choose to investigate this factor?

_____

_____

_____

2. Which environment did the worms prefer? Did the worms' behavior support your hypothesis?

_____

_____

_____

3. What is it about an earthworm's structure and requirements for life that explain their responses to your environmental conditions?

_____

_____

_____

4. Describe the variables that were controlled in your experiment. Why is it important that these variables remain constant?

_____

_____

_____

_____

5. Error Analysis What were some possible sources of error in your experiment?

_____

_____

_____

_____

6. Exchange your experimental design and results with another group. What do their data indicate about the general behavior of earthworms?

_____

_____

_____

_____

Write and Discuss
Write a short paragraph describing your findings and indicating whether or not they support your hypothesis. Did both of the worms you studied behave in the same manner? Discuss any questions your results might have raised.

_____

_____

_____

_____

Inquiry Extensions
1. What other earthworm behaviors could you investigate? Develop a hypothesis and a procedure for testing it. What would you expect to see?

2. Develop a hypothesis about the role of earthworms in the biological niche they inhabit. Write an experimental procedure to test your hypothesis. If your teacher approves of your plan, conduct the experiment and report your results.

Design Your Own
Lab 28

What is living in the leaf litter?

Have you ever walked in the woods and wondered what creatures were living in the soil beneath you? How do these organisms impact the environment, and what role do they play in the food web?

Many organisms live in the soil and leaf litter above the soil. These organisms play a role in the health of the habitat and can impact the environment in both positive and negative ways. Arthropods and other organisms in the leaf litter feed on items found there, as well as on each other. In turn, other animals feed on them. In this experiment, you will discover what types of organisms live in the soil in your area and infer the role these organisms play in the soil's food web.

Problem
A species of bird that is strictly insectivorous has changed its migration path. Determine what food is available to the birds.

Objectives
- Observe organisms found in soil or leaf litter.
- Identify the organisms.

Safety Precautions
🥽 👔 🔬 🧤 🦺

WARNING: *Use care when handling the scissors and wire screening; their edges are sharp and might cut or puncture the skin.*

Possible Materials
2-L clear plastic bottle
scissors
trowel
cheesecloth or plastic wrap
rubber bands
desk lamp
magnifying lens
jars—one large, one small
1/4-inch mesh wire screen (10-cm square)
forceps
spoons
500–1000 mL soil sample
leaf litter
pine cones
identification guide

Hypothesis
Use what you know about arthropods to write a hypothesis about their presence in the soil environment.

_____

_____

_____

_____

Design Your Own Lab 28, What is living in the leaf litter? continued

Plan the Experiment

1. Read and complete the lab safety form.
2. Choose one of the following substances to test for arthropods: leaf litter, dry sandy soil, soil near a pond, or pine cones and pine needles from a forest. Which type of substance do you think will have a varied and diverse population of arthropods?
3. Decide on a procedure to collect and examine the arthropods from the soil or leaf sample. In the space provided, write your procedure for collecting and separating the arthropods. Include the materials you will use.
4. Decide how you will record your data. Create a data table to hold your observations. Include room for quantitative and qualitative data on the organisms as well as detailed sketches.

Check the Plan

1. Make sure your teacher has approved your experimental design before you begin.
2. Do not handle the animals; they might bite or sting.
3. When you have completed your experiment, dispose of the materials as directed by your teacher.

Record the Plan

In the space below, write your experimental procedure and make a sketch of your experimental setup.

Data and Observations

1. Use the space below to create a data table of your findings, including a sketch of the organisms, the number found, the size, important characteristics, and a preliminary identification of each organism.

Design Your Own **Lab 28, What is living in the leaf litter?** continued

2. In the space below, provide detailed sketches of several of the arthropods you found and label the body parts.

Analyze and Conclude

1. How did the conditions of your soil sample change from the beginning of the lab to the end of the lab? What caused these changes?

_____

_____

_____

2. What types of animals did you find in your sample? How did you identify these organisms?

_____

_____

_____

_____

3. What factor or factors made the animals in your sample move?

_____

_____

_____

_____

4. What was the primary method of locomotion for the animals you found? Were there any exceptions? If there were different means of locomotion available to the animal under observation, draw conclusions about how the animal would use that means in the environment.

_____

_____

_____

_____

5. Error Analysis What were some possible sources of error in your lab design and observations?

_____

_____

_____

6. Exchange your procedure and data with another group in your class for peer review. What do their data indicate about the presence of arthropods in the soil samples they were using?

_____

_____

_____

Write and Discuss

Write a short paragraph describing your findings and indicating whether or not they support your hypothesis.

_____

_____

_____

_____

_____

Inquiry Extensions

1. What other organisms, besides the arthropods you examined, could be present in your soil? What relationship might exist between the arthropods and these organisms?

2. How do arthropods survive the winter? What do they do if or when the soil freezes? Develop a hypothesis that would explain the effects of the seasons on these organisms and design an experiment to test it.

Classic
Lab 29

How can you analyze echinoderm relationships?

A cladogram is a diagram which shows the evolutionary relationships between different groups of organisms. By showing these relationships, cladograms essentially reconstruct the evolutionary history (or phylogeny) of these organisms. Cladograms are also sometimes known as phylogenies or phylogenic trees.

Cladograms are constructed by grouping organisms together based on their shared characteristics. Scientists collect data on the features of all the organisms they want to classify. This includes physical features, such as whether or not the animal is a vertebrate or has limbs, as well as physiological features, such as how it obtains nourishment. Scientists then analyze this data to determine which characteristics were present in common ancestors and which could have developed at a later time.

If you were to study a group of organisms, you would most likely find characteristics that all members of the group share. These are known as primitive, or original, characteristics. Characteristics shared by only a portion of the group are considered derived characteristics. Scientists believe that these are more advanced features, or adaptations, that helped the organisms survive. These derived characteristics are the basis of constructing cladograms.

In this lab, you will design your own cladograms based on the characteristics of several animal specimens. Some representative echinoderm species are shown in **Figure 1**.

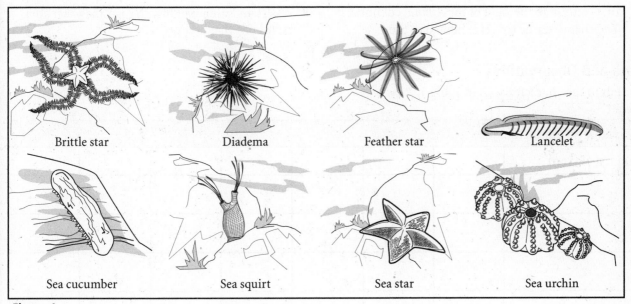

Brittle star   Diadema   Feather star   Lancelet

Sea cucumber   Sea squirt   Sea star   Sea urchin

Figure 1

Objectives

- Examine samples of echinoderms and invertebrate chordates to determine characteristics.
- Create a cladogram to represent the evolutionary relationships among these animals.
- Make inferences about the significance of these relationships.

Materials

diagrams
ruler
glue
paper
markers/colored pencils
specimens: sea star, brittle star, sea urchin, sea cucumber, feather star, sea squirt, and lancelets

Classic **Lab 29, How can you analyze echinoderm relationships?** continued

Safety Precautions

WARNING: *Sea urchin spines are very sharp and can puncture the skin.*

Procedure

1. Read and complete the lab safety form.
2. Obtain a set of echinoderms and invertebrate chordates. Lay them out on the lab table. If you are observing living specimens, handle them gently. Do not harm the animals in any way.
3. List all the characteristics you see for each animal, and assign each one to a column in the top row of **Table 1**.
4. The taxon (a group of animals or plants that share most of their characteristics) with the least number of derived characteristics should be listed in the first row. The taxon with the greatest number of derived characteristics should be listed in the last row. The characteristic that is evident in the greatest number of taxa should be listed in the first column to the right of the *Animal* column. The characteristic that is exhibited in the least number of taxa should be listed in the last column. Use plus and minus symbols to represent presence (+) or absence (−) of specific characteristics in species.
5. Determine which characteristic all the organisms have in common.
6. Look at the data, and determine the derived characteristics. The largest group of these derived characteristics will be the first to branch from the main trunk of the cladogram. Name the derived characteristic, and list all the animals that have that characteristic.
7. Look for other characteristics that are common to only a portion of the group, and add these to the cladogram until the groups can be sorted no further.
8. Wash hands thoroughly with soap and water after examining the specimens.

Data and Observations

1. Fill in the characteristics in **Table 1**.

Table 1

Echinoderm Characteristics							
Animal (Taxon)							

Classic **Lab 29, How can you analyze echinoderm relationships?** continued

2. In the space below, create your cladogram of echinoderms and invertebrate chordates.

Analyze and Conclude

1. What was the primitive characteristic for the animals you examined? Why was this characteristic of little use when designing your cladogram?

_____

_____

_____

2. What characteristics did you choose to examine? Why did you choose these?

_____

_____

_____

3. What did the characteristics you chose tell you about the way that the animal moves or eats?

_____

_____

_____

4. What is the main difference between echinoderms and invertebrate chordates?

_____

_____

_____

_____

5. Error Analysis What were some possible sources of error in your analysis?

_____

_____

_____

6. Which organism is most closely related to sea stars? Which is most distantly related? Explain your answer.

_____

_____

_____

_____

_____

Inquiry Extensions

1. An estimated 6000 species of echinoderms and about 1200 known species of invertebrate chordates exist. What is the evolutionary history of these species? Find examples of each group that existed long ago, and show how these animals have changed throughout their evolutionary history. Create a diagram that shows this transition, and include the time periods of each.

2. How do echinoderms and invertebrate chordates move and eat? Use a variety of materials to create a model of one organism that shows an animal's mobility and consumption methods. Share your models with the class.

How have frogs adapted to land and aquatic habitats?

Amphibians are the evolutionary bridge between fishes and reptiles. Frogs exemplify most of the traits that characterize amphibians. In this lab, you will observe an adult frog and identify those traits that make it suited to a terrestrial environment.

Objectives
- Observe a live frog.
- Compare and contrast a frog to humans.
- Identify useful adaptations of frogs.
- Research one adaptation.
- Make a model of the adaptation to share with the class.

Materials
live frog
paper towels
tabletop
water
pencil with an eraser
large aquarium

Safety Precautions

Procedure
Part A. Observing the Frog
1. Read and complete the lab safety form.
2. Moisten the top of the table where the frog will be placed. Sit quietly by the table, and allow time for the frog to become accustomed to its surroundings. By avoiding sudden movements, you will increase your opportunities for making accurate observations.
3. Compare the general structure of the frog's body with that of your own. Think of your body as consisting of a head, neck, trunk, and four appendages. Record your observations in **Table 1**.
4. Locate the frog's eyes. The ears are located behind and below the eyes. The eardrum is stretched across the ear opening.
5. In the human body, each of the upper appendages consists of a series of parts called the upper arm, the forearm, the wrist, the hand, and the fingers; each of the lower appendages consists of the thigh, shank, ankle, foot and toes. Identify similar structures, if possible, in the frog. Record your observations in **Table 1**.
6. Using the eraser end of a pencil, gently prod the frog until it jumps.
7. You must observe carefully to see the frog breathe. First locate the nostrils. (Ducts lead from the nostrils to the posterior part of the mouth cavity.) Then, without touching the frog, watch the floor of the mouth (upper throat). When it is lowered, the mouth cavity enlarges.
8. Place the frog in the water at one end of the large aquarium. Observe the motion used to swim.
9. Return the live frog to the container designated by your teacher.
10. Make a diagram of a frog, and label the parts that you saw in your observations. Identify your frog as male or female.

Part B. Amphibian Adaptation

1. Look back at the observations you made of the live frog. As you have seen, amphibians show many adaptations that allow them to spend part of their life cycle in an aquatic environment and part of their life cycle in a terrestrial environment.

2. Choose one of these adaptations to study. Use resources to explore more about this particular adaptation and how it supports the amphibian in a terrestrial environment.

3. Design an interactive display or demonstration that will show your classmates why this particular adaptation is useful. Tie your demonstration to evolutionary pressures, and describe how this adaptation has provided an evolutionary advantage for the frog.

4. As part of a science symposium, present your demonstration or display to the class.

Data and Observations
Table 1

Observation Data		
Trait	**Frog**	**Human**
Body shape/length		
Neck		
Eyes		
Ears		
Nostrils		
Skin		
Feet		
Length of hip to knee and knee to foot		
Movement type/length of stride		
Breathing		

Classic **Lab 30, How have frogs adapted to land and aquatic habitats?** continued

Analyze and Conclude

1. How is the physical structure of a frog similar to that of a human? How is it different? What type of symmetry does a frog's body have?

_____

_____

_____

_____

2. How does a frog breathe?

_____

_____

_____

_____

3. Describe a frog's eyes. How are they different from yours? How are they the same?

_____

_____

_____

4. What adaptations make a frog suited to its life in the water and on land? Explain.

_____

_____

_____

_____

5. Which adaptation did you choose to further explore? Why did you choose this one?

_____

_____

_____

_____

_____

_____

6. What conclusions did you draw from your research? Do all amphibians show the same adaptation?

_____

_____

_____

_____

7. Error Analysis What errors could have been made while observing the live frog?

_____

_____

_____

_____

Inquiry Extensions

1. Choose another amphibian, such as a salamander or newt, to observe. Compare the structures and behaviors you notice with those of the frog.

2. Many amphibians have some means of protection. Some are camouflaged, while some have a poison in their skin. Research one such adaptation, and create a short information pamphlet about it to share with your class.

Classic
Lab 31

What are the structures and functions of a chicken egg?

Fossil evidence suggests that birds evolved from archosaurs—the line from which crocodiles and dinosaurs evolved. This indicates that dinosaurs, crocodiles, and birds are more closely related to each other than they are to turtles and lizards.

A chicken egg is made up of an egg yolk, albumen, egg membranes, and a shell. Eggshells are semipermeable. They allow gas to move in and out of the egg but keep out most liquids. Materials move in and out of the egg by osmosis—the movement of water molecules from an area of high concentration to an area of low concentration.

In this lab, you will examine chicken eggs to explore how their structures have adapted to an existence on land.

Objectives
- Identify the different parts of a chicken egg.
- Recognize how substances can pass across a membrane during osmosis.
- Infer how the egg is suited to existing on land, as opposed to in the water.

Materials
petri dish
tweezers
magnifying lens
metric ruler
2 unfertilized raw chicken eggs (the larger the better)
hard-boiled chicken egg (the larger the better)
paper towels
small plastic knife
microscope
microscope slide
cover slip
distilled water
clear corn syrup
vinegar
clear plastic cup
balance
rubber gloves
aluminum foil

Safety Precautions

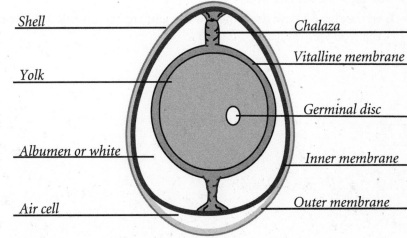

Figure 1

Procedure

Part A. Structures of an Egg

1. Read and complete the lab safety form.
2. Put on a pair of rubber gloves, and obtain a raw egg from your teacher.
3. Carefully crack and open the egg, and place it in the petri dish. If possible, try to break the egg in half and keep as much of the yolk and white in one half of the shell. Try not to break the shell into many pieces.
4. Obtain a hard-boiled egg from your teacher. Carefully remove the shell, and cut through the hard-boiled egg with the plastic knife. Lay both halves of the egg and the broken shell on the paper towel.
5. Compare the two eggs to **Figure 1**. In the space in the *Data and Observations* section, draw and label the different parts of each egg.
6. The germinal disk is a white dot in the center of a raw egg yolk. Use a ruler to measure the diameter of that dot.
7. Compare the yolks, the albumen, and the shells of the hard-boiled egg and the raw egg using the microscope. Record your observations in **Table 1**.

Part B. Osmosis

1. Choose a raw egg, and gently wipe it clean with a paper towel. Place it on the balance, and measure the mass in grams. Record this information in **Table 2**.
2. Place the egg in a clear plastic cup, and cover it with 150 mL of vinegar.
3. Cover the cup with aluminum foil, and leave it undisturbed for two days.
4. After two days, the eggshell should have dissolved away. Wearing gloves, gently remove the egg from the beaker. Rinse it under running water, then gently dry it with a paper towel.
5. Measure the mass of the egg again, and record the mass in the data table. Also record any changes that might have occurred to the egg.
6. Measure the amount of vinegar left in the beaker, and record this information.
7. Clean and dry the cup, and place the egg back in it. Add 150 mL of clear corn syrup to the cup.
8. Cover the cup and leave undisturbed for 24 h.
9. Again wearing gloves, carefully remove the egg, and rinse it under running water.
10. Measure the mass of the egg, and record the mass and any other changes in the data table.
11. Measure the amount of corn syrup left in the beaker, and record this information in the data table.
12. Clean and dry the cup, and place the egg back in it. Add 150 mL of distilled water.
13. Repeat steps 8–11 for the egg and the distilled water.
14. Dispose of the egg as directed by your teacher.

Data and Observations

1. In the space below, draw the raw egg you are examining. Label the different parts that you see.

Classic **Lab 31, What are the structures and functions of a chicken egg?** continued

2. In the space below, draw a diagram of the hard-boiled egg you are examining. Label the visible parts.

Table 1

Comparison of Egg Structures					
	Shell	**Yolk**	**Albumen**	**Membranes**	**Chalazae**
Hard-boiled egg					
Raw egg					

Table 2

Osmosis in Shelled Egg						
Liquid	**Amount of Liquid**		**Mass of Egg**		**Egg Observations**	
	Before	**After**	**Before**	**After**	**Before**	**After**
Vinegar						
Corn syrup						
Distilled water						

Analyze and Conclude

1. Is the eggshell porous? Is it permeable to air? To water? How is the eggshell of a hard-boiled egg different from the shell of a raw egg?

_____

_____

_____

2. Which part of the egg is analogous to the placenta in a mammal?

_____

_____

3. How did the egg change when it was put in vinegar? Why did this occur?

_____

_____

4. How did the egg change when it was put in corn syrup? Why did this occur?

_____

_____

5. Describe the changes in the egg when it was placed in distilled water. Explain what you saw.

_____

_____

_____

6. Error Analysis What were some possible sources of error in your experiment?

_____

_____

_____

7. How is a chicken egg suited for life on land? How do you think an amphibian egg might be different, considering amphibian eggs are laid in water?

_____

_____

_____

Inquiry Extensions

1. All birds are oviparous—the young grow in eggs outside the mother's body. Some animal groups, such as fishes and reptiles, include genera that are oviparous and some that are ovoviviparous. Research what this term means. Cite some animals that fall into this category. What advantage might this give to the organisms that use it?

2. Many dinosaurs laid eggs. Research the egg-laying behavior of at least two species of dinosaur, and compare the dinosaur behavior to that of birds and reptiles. Present your findings to the class.

Design Your Own
Lab 32

What is the best way to keep warm?

Many mammals and other warm-blooded animals have some sort of insulation on their body to help them maintain a constant body temperature despite variations in the temperature surrounding them. For example, a whale has a thick layer of blubber, and a sheep has a thick wool coat. In this lab, you will model the insulating properties of wool.

Problem

Many people wear wool clothes in the winter to keep warm. How well does wool insulate when it gets wet?

Objectives

- Form a hypothesis comparing the insulating properties of wet wool socks and dry wool socks.
- Design an experiment to test the hypothesis.
- Compare the temperature of water in the socks over time.

Safety Precautions

Possible Materials

wool socks (1 pair)
1-L glass beaker
plastic containers with lids (3)
thermometers (non-mercury) (3)
hot tap water
room-temperature water
stopwatch
scissors
craft knife
rubber bands
graph paper
colored pencils

Hypothesis

Use what you know about wool as an insulating material to write a hypothesis indicating the differences between the insulating ability of dry wool socks and wet wool socks.

_____

_____

_____

_____

Design Your Own **Lab** 32, **What is the best way to keep warm?** continued

Plan the Experiment

1. Read and complete the lab safety form.
2. Determine how you will test the insulating properties of a wet wool sock and a dry wool sock. Assemble your experiment.
3. Identify the independent variable, dependent variable, constants, and control for your experiment.
4. Decide how you will record your data and when you will record it. Design and construct a data table to hold your experimental data on temperature, container type, and time.
5. Decide how you will present your findings to the rest of the class.

Check the Plan

1. Make sure your teacher has approved your experimental plan before you proceed.
2. Be sure that a control is included in your experiment and that the experimental groups vary in only one way.
3. **Figure 1** Use caution preparing the container lids and when pouring hot water into the containers.
4. Observe the changes in temperature in containers in wet and dry socks.
5. When you have completed the experiment, clean up your materials as directed by your teacher. Remember to wash hands with soap and water after completing the lab.

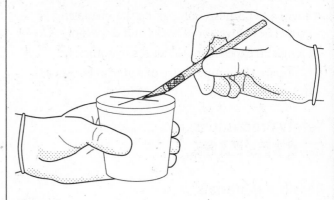

Figure 1

Record the Plan

In the space below, write your experimental procedure and make a sketch of your experimental setup.

Design Your Own **Lab 32, What is the best way to keep warm?** continued

Data and Observations

1. Use the space below to create a data table of your findings, including the container, temperature, and time lapsed.

2. In the space below, or on a separate sheet of graph paper, graph the results of your experiment.

Design Your Own **Lab** 32, **What is the best way to keep warm?** continued

Analyze and Conclude

1. How did the temperatures change in your containers?

_____

_____

2. Describe the control in your experiment. What did the control show?

_____

_____

3. What comparisons can be made between the wool socks in your experiment and the insulation that some mammals have? Use specific examples in your answer.

_____

_____

_____

4. **Error Analysis** What were some possible sources of error in your experiment?

_____

_____

5. Some manufacturers claim that their wool socks will keep you warm, even when wet. Do your findings support this claim? Explain.

_____

_____

6. Exchange your procedure and data with another group for peer review. What do their data indicate about the insulating power of wool?

_____

_____

_____

Write and Discuss

On a separate sheet of paper, write a short paragraph describing your findings and indicating whether or not they support your hypothesis. Was there a difference between the insulating ability of wet wool and dry wool? Discuss any questions your results might have raised.

Inquiry Extensions

1. How would you design an experiment to compare the insulating properties of wool with those of another fabric, such as cotton or rayon, or other insulating materials? What would you do differently?

2. Research the insulation of animals that live in water and animals that live on land. What types of tissues keep them warm?

Classic Lab 33

How do we learn?

You have been a student for long enough to know that some things are easier to learn than others. What happens when we are presented with a new task? What helps people remember new tasks? What interferes with performance?

In this lab, you will design a test that examines the way a subject masters a new task.

Objectives

- Conduct an experiment to answer questions about human learning.
- Make predictions about learning and performance.
- Communicate your findings in an appropriate manner.

Materials

pencil or marker
maze puzzle (12 copies)
portable CD or MP3 player with earphones
music or voice recording
stopwatch

Procedure

1. Read and complete the lab safety form.
2. Work with a partner. Before you begin the experiment, read through the procedure and make a prediction about the effects of the conditions.
3. **Figure 1** Set up a maze race exercise, as shown below. Designate one person as the racer and the other as the timer.
4. Have the racer complete the maze while looking only at the reflection of his or her hands in the mirror.

Figure 1

5. Use the stopwatch to time the racer's performance, and record the data in **Table 1**. Repeat this exercise twice more with fresh copies of the maze. This will complete trials 1–3.

6. For trials 4–6, repeat the exercise with a new copy of the same maze, but this time have the racer listen to music or a voice recording while performing the task.

7. Record the time for the second set of trials in **Table 1**.

8. For trials 7–12, repeat steps 4–7, but do not use the mirror. Have the racer look directly at the paper. Record the results in **Table 2**.

Data and Observations

Table 1

Trial	Results With Mirror
1	
2	
3	
4	
5	
6	

Table 2

Trial	Results Without Mirror
7	
8	
9	
10	
11	
12	

Classic **Lab 33, How do we learn?** continued

Analyze and Conclude

1. Did the racer's performance support your prediction? Explain.

_____

_____

_____

2. By the end of trial 12, was there overall improvement in completing the task? Explain.

_____

_____

_____

3. Based on your observations and data, how did the racer perform when completing the task with the stressor (the distraction)? Explain.

_____

_____

_____

4. Think of an animal learning a new behavior. What advantage might there be for an animal that can learn under stress?

_____

_____

_____

5. Error Analysis What were some possible sources of error in your experiment?

_____

_____

_____

_____

6. Share your data with other students. Discuss how the data of other groups compares with yours.

_____

_____

_____

_____

7. Use the space below to make a line graph showing the results of your experiment. Using a different color, draw an additional line to show your prediction of the results if trials 7–9 were repeated two or three more times immediately following trial 12. What would you expect to see?

Inquiry Extensions

1. How does time impact memory? Create an experiment that investigates how long people remember what they learn.

2. Have you ever noticed that you can remember all the words to your favorite songs but cannot remember the formulas for respiration or photosynthesis? Take a difficult concept you have learned in biology, and write it as lyrics to a familiar tune that makes it easy to remember. Present your song to the class.

Classic
Lab 34

How long can you last?

Your body has three types of muscles: smooth, cardiac, and skeletal. Muscle cells can only contract or relax, so skeletal muscle groups must work in pairs. When one muscle group contracts, the opposing muscle group relaxes to its original shape. For example, when you bend your elbow, the biceps group on your upper arm contracts. As the biceps on the front of your arm contract, the triceps on the back of your arm relax to their original length. When you want to straighten your elbow, the process works in reverse. The triceps contract while the biceps relax.

In this lab, you will be examining the way that muscles work together in your body to squeeze a rubber bulb. You will be working to see what impact fatigue and repetitive work has on the muscles in your hand, arm, and leg.

Objectives
- Assemble a muscle-testing device based on directions and diagrams.
- Test the muscle fatigue experienced after repetitive motion over a period of one minute.
- Serve as a counter and recorder as other group members test muscle fatigue.
- Draw conclusions about muscle fatigue.

Materials
rubber bulb
1/2-inch plastic or vinyl hose, 1 m in length
1/2-inch PVC pipe, 20 cm in length
ring stand
paper flag, 2.5 cm square; red on one side, white on the other
stopwatch
cellophane tape
duct tape
first aid tape

Safety Precautions

WARNING: *Do not pull on the tubing—the apparatus might become unstable and fall over.*

Procedure
Part A. Assembling the Device
1. Read and complete the lab safety form.
2. In your group, assemble the muscle fatigue device as indicated in **Figures 1** and **2**. Attach a small rubber bulb to the end of a hose. (If necessary, cut a half-inch slit in the hose to allow enough room for the nozzle of the bulb.) The hose should then be inserted into the small length of PVC pipe and the entire device held up by a ring stand as shown in **Figure 1**. Use duct tape to secure both connections.
3. Tape a small paper flag over the top of the pipe so that it can move freely but will still reset itself. Tape it so that the white side is facing up.

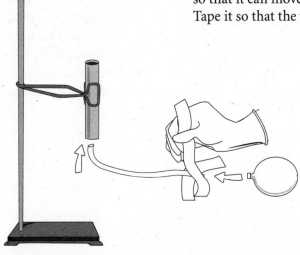

Figure 1

4. Test your device by squeezing the bulb, as shown in **Figure 2**. The air should lift the flag, making the red side visible, as shown in **Figure 3**. When the bulb is released, the flag should return to its starting position.

5. Decide which group member will act as Observer 1, Observer 2, and Tester. You will be rotating positions as time allows.

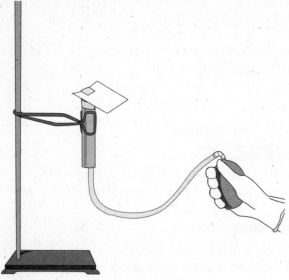

Figure 2

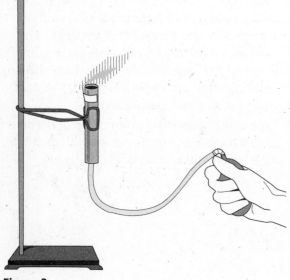

Figure 3

Part B. Testing for Fatigue

1. Complete the data table on the next page to hold the observations and data gathered in this experiment. The data table should have room for the names of each group member acting as the Tester, muscle group tested, the number of compressions in one minute, the number of time the flag lifts in one minute, and any comments the tester had after each test.

2. Start with the first Tester. This person will place the bulb in his or her hand. When Observer 1 gives the go-ahead, the Tester squeezes the bulb as many times as he or she can in the span of one minute.

3. Observer 1 has the job of timing one minute and also of counting the number of times the Tester squeezes the bulb. This information should be put in **Table 1**.

4. Observer 2 should position his or her eyes at the same level as the top of the tube, then count the number of times the red side of the flag appears during the one-minute trial. This information should be recorded in **Table 1**. The Tester should also insert any comments he or she has about muscle fatigue in **Table 1** at this point. After allowing a 15-s rest, repeat steps 2-4 for a second trial.

5. Keeping the same roles, the Tester will now place the bulb between the forearm and the upper arm (on the inside of the elbow). This will test his or her arm muscles.

6. Conduct the test for one minute, with Observer 1 and Observer 2 watching and counting the number of squeezes and the number of times the red flag appears. Repeat this test for a second trial.

7. The same Tester should then place the bulb behind the knee to test his or her leg muscles. Conduct two trials.

8. Rotate roles so that each member of the group performs all three tests.

9. Follow the instructions given by your teacher about disassembling your device.

Classic **Lab** 34, **How long can you last?** continued

Data and Observations
Table 1

Fatigue Data						
		Trial 1		Trial 2		Comments
		Number of Squeezes	Number of Red Flags	Number of Squeezes	Number of Red Flags	
Tester 1	Hand					
	Arm					
	Leg					
Tester 2	Hand					
	Arm					
	Leg					
Tester 3	Hand					
	Arm					
	Leg					

Analyze and Conclude

1. Describe the motion of the arm muscles, leg muscles, and hand muscles as the bulb was squeezed.

_____

_____

_____

_____

2. What is muscle fatigue? Describe the muscle fatigue you found in this exercise.

_____

_____

_____

_____

3. Which movement was easiest for you? Which was most difficult? Why do you think that may have been?

_____

_____

_____

4. Which muscle group became fatigued the quickest? Which was least fatigued after the one-minute trial?

_____

_____

_____

5. Error Analysis What were some possible sources of error in your experiment?

_____

_____

_____

_____

6. How did your results compare with those of your teammates? Can you think of reasons why there might be differences?

_____

_____

_____

_____

Inquiry Extensions

1. How does resting the muscles between trials impact the results? Conduct the experiment again but allow each Tester the opportunity to rest for 30 s before resuming with the same muscle group. What differences, if any, are there in the results? Explain the differences.

2. Is there another way to test the endurance of these muscle groups? Conduct an experiment in which you time isometric muscle contractions (for example, by standing on one leg, or by using your arm to hold your body at an angle against a wall or table) for one minute, then repeat the trial after a 30-s rest. At what point(s) were the muscles fatigued and unable to continue their work? Do the Testers report that this test was easier or harder than the bulb test? What conclusions can you draw about continuous muscle contractions as opposed to repetitive contractions?

Design Your Own
Lab 35

How quickly do you respond?

Your nervous system receives information about what is happening both inside and outside your body. Any change or signal in the environment that can make an organism react is called a stimulus. Your nervous system will analyze the stimulus and initiate a response or a reaction. It helps you move, think, feel pain, and enjoy a chocolate-chip cookie. Your nervous system also plays a role in maintaining homeostasis and recognizing basic needs of survival such as oxygen, water, and nutrients.

Problem

Design a lab experiment that measures driver reaction time for you and your lab partners. *Under no circumstances should any experimental plan be carried out in a real car.*

Objectives

- Form a hypothesis about the effect that one variable, such as time of day, has on reaction time.
- Design an experiment to test the effect of your variable on hand and foot reaction times.
- Compare the results of reaction times with and without the variable.

Safety Precautions

Possible Materials

stopwatch
cover of frying pan (for steering wheel)
small blocks of wood (for brake and gas pedals) (2)
rubber mat
rubber ball

Hypothesis

Use what you know about the nervous system and your own response time to write a hypothesis indicating how a driver's reaction time is affected by a variable you determine.

_____

_____

_____

_____

Design Your Own **Lab** 35, **How quickly do you respond?** continued

Plan the Experiment

1. Read and complete the lab safety form.
2. Choose a way to test eye-to-foot reaction time and eye-to-hand reaction time.
3. Decide on your procedure for collecting your data. In the space provided, write your procedure for testing the reaction times of your lab partners. Include the materials you will use.
4. Identify the independent variable, dependent variable, constants, and control group.
5. Decide how you will record your data and when you will record it. Design a data table to collect information.

Check the Plan

1. Make sure your teacher has approved your experimental plan before you proceed.
2. Be sure that a control is included in your experiment.

Record the Plan

In the space below, write your experimental procedure and make a sketch of your experimental setup.

Design Your Own **Lab** 35, **How quickly do you respond?** continued

Data and Observations

1. Use the space below to create a data table of your findings, including the type of reaction being tested, the reaction time of each person, and the length of time that has passed.

Analyze and Conclude

1. What relationship did you find between reaction time and the variable you selected? Explain.

_____

_____

_____

2. Based on your observations, what recommendations would you make for teenage drivers? Explain.

_____

_____

_____

3. Did you see any differences between the reaction time between eyes to hand and eyes to foot? Explain.

_____

_____

_____

4. Describe the control in your experiment. What did the control show?

_____

_____

5. Error Analysis What were some possible sources of error in your experiment?

_____

_____

_____

6. Exchange your procedure and data with another group in your class for peer review. What do their data indicate about reaction times?

_____

_____

_____

Write and Discuss

Write a short paragraph describing your findings and indicating whether or not they support your hypothesis. Discuss any questions your results might have raised.

_____

_____

_____

_____

Inquiry Extensions

1. Cell phone use while driving is a controversial issue. Many people now use hands-free devices on their phones so they can pay more attention to the road, but some people question the safety of these devices as well. Design an experiment that compares the reaction time of someone who is using a hand-held phone and someone who is using a hands-free device. *Under no circumstances should any experiment be carried out in a real car.* Why do you think some states fine drivers for talking on cell phones? Relate this to your experimental data.

2. When does stimulus turn into stress? Generally, a stimulus can be a single, simple event such as an itch or stubbing one's toe. Stress, on the other hand, tends to describe a set of stimuli that prompt more intense, and sometimes long-lasting, physiological and emotional responses. (Negative stressors cause *distress*, while positive stressors cause *eustress*.) Devise a journal to track a few hours of a given day. Record a selection of stimuli and your responses to them, then characterize each stimulus. Were they simple stimuli or stressors? What was your recovery time? Are the stressors acute or chronic? When you have completed gathering the data, create a chart or graph to show duration and intensity of the selected events and your responses.

Design Your Own
Lab 36

How much air can your lungs hold?

Every day you breathe in and out thousands of times. How much air do you take into your lungs each time? What conditions influence your lung capacity? Imagine that a friend who is close to you in age will be coming to your area to participate in a charity walk-a-thon with you. The air temperature in your friend's hometown is dramatically different from the typical temperatures in your area at this time of year. Will this temperature difference affect your friend's lung capacity?

Problem
Lung capacity can be affected by environmental factors.

Objectives
- Identify a factor that influences lung capacity.
- Design an experiment to test that factor.
- Draw conclusions about what impacts lung capacity.

Safety Precautions

WARNING: *Do not share balloons with other classmates. Do not put balloons or broken pieces of a balloon in the mouth; this presents a choking hazard. Students who have asthma, breathing difficulties, or a latex allergy should discuss their participation with the teacher.*

Possible Materials
round balloons (12-inch maximum diameter)
thermometer (non-mercury)
measuring tape
string
metric ruler
calculator

Hypothesis
Use what you know about lung capacity to write a hypothesis that predicts the impact air temperature has on lung capacity.

_____

_____

_____

_____

Plan the Experiment

1. Read and complete the lab safety form.
2. Make a list of the factors that might influence the lung capacity of the average high school student. Be sure to mention the factors cited in your hypothesis.
3. Decide on a procedure for testing your hypothesis. In the space provided, write your procedure for testing lung capacity. Include a list of the materials you will use.
4. Identify the independent variable, dependent variable, constants, and control group.
5. Decide how you will record your data and when you will record it. Design a data table to collect information about change in lung capacity in cubic centimeters.

Check the Plan

1. Be sure that a control group is included in your experiment and that the experimental groups vary in only one way.
2. Make sure your teacher has approved your experimental plan before you proceed.
3. When you have completed your experiment, dispose of materials as directed by your teacher.

Record the Plan

In the space below, write your experimental procedure and make a sketch of your experimental setup.

Design Your Own **Lab 36, How much air can your lungs hold?** continued

Data and Observations

1. In the space below, create a data table to hold the information gathered in this experiment.

Analyze and Conclude

1. Why might it be important to know a person's lung capacity?

_____

_____

_____

2. What did you learn from your experiment about the factors that you studied?

_____

_____

_____

3. How does your lung capacity differ from that of other students in your class? What factors might account for these differences?

_____

_____

_____

_____

4. How might you design your experiment differently next time?

_____

_____

_____

5. **Error Analysis** What were some possible sources of error in your experiment?

_____

_____

_____

_____

6. Exchange your procedure and data with another group in your class for peer review. What do their data indicate about the impact of different factors on lung capacity?

_____

_____

_____

_____

Write and Discuss

Write a short paragraph describing your findings and indicating whether or not they support your hypothesis. Discuss any questions your results might have raised.

_____

_____

_____

_____

Inquiry Extensions

1. What differences would you expect to see in the lung capacity of smokers as compared to nonsmokers, and as compared to nonsmokers who live with smokers? Design a questionnaire for the participants in your study to determine their exposure to cigarette smoke and other factors that can affect lung capacity. Then design an experiment to test their lung capacities. Share your findings with the rest of your class.

2. What other environmental factors, such as smog or altitude, impact lung capacity? Are there certain professions or chronic diseases that cause decreased lung capacity? Design an experiment to study these factors. Keep in mind that you might not be able to conduct your experiment due to geographic location or season, but predict what changes you might see. Follow up with research to find data on the factors or populations you chose to study.

Classic
Lab 37

How healthy are they?

When people visit the doctor for a check up they usually have a urine sample taken. A urine sample can often be useful in detecting medical conditions that might not show other symptoms. Normal, healthy urine contains little or no glucose or protein. Glucose in the urine might be an indication that the person has diabetes. Diabetes is a disease where the body cannot use enough glucose from the blood. If there is protein in the urine, this might be a sign that the kidneys are not functioning properly. In this lab, you will be testing simulated urine samples to try to diagnose diabetes or kidney diseases.

Objectives
- Test simulated urine samples for the presence of glucose using glucose test strips.
- Test simulated urine samples for the presence of protein using Biuret solution.
- Create a data table.
- Interpret test results.

Materials
test tubes (6)
test-tube rack
plastic droppers (6)
water
glucose solution
protein solution
glucose test strips (6)
large pipettes for each sample
Biuret solution
simulated urine samples (3)
wax pencil
white paper or paper towels

Safety Precautions

WARNING: *Use caution with the Biuret solution. It is corrosive. It can irritate the eyes, skin, or respiratory tract, and should NOT be ingested. If contact occurs, flush the affected area with cold water. It might also stain clothing.*

Procedure
Part A. Test for Glucose
1. Read and complete the lab safety form.
2. Create a data table that will hold the data you gather during the course of this lab. There is a sample data table included in the next section.
3. Label six test tubes in the following way: *G* for glucose, *P* for protein, *W* for water and *1, 2*, and *3* for the three simulated urine samples.
4. **Figure 1** shows you how to place the test tubes in the test-tube rack.
5. Obtain six glucose strips and label them in the same fashion: *G, P, W, 1, 2*, and *3*.

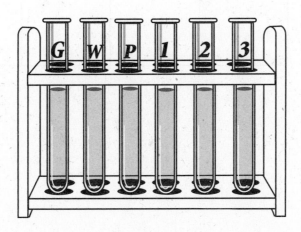

Figure 1

6. Using a large pipette, fill each test tube about ¾ full with the appropriate solution.

7. Place the *G* test strip on the clean paper towel. Use a clean dropper to pick up some of the liquid from the *G* test tube.

8. Figure 2 shows you how to drop 2 drops of the glucose solution on the glucose test strip.

9. Record any change of color in your table. If there is no change, write "No change".

10. Repeat steps 7–9 for each of the solutions, recording your observations in your data table. Be sure to complete Part A before continuing with Part B.

Part B. Test for Protein

1. Get a bottle of Biuret solution from your teacher. Use caution: Biuret solution is corrosive and can irritate the skin, eyes, and respiratory tract. If contact occurs, flush the affected area with cold water. The solution can also stain clothing.

2. Take note of the original color of the solution.

3. Carefully add 30 drops of the solution to the test tube labeled *G*.

4. Gently swirl the test tube to mix the liquids.

5. Note any color change. You might want to hold the tube in front of a white paper towel to provide a neutral background.

6. Observe any changes and record the changes in **Table 1**.

7. Repeat steps 3–5 for the remaining solutions.

8. Clean the equipment and dispose of the liquids as directed by your teacher. Wash your hands with soap and water.

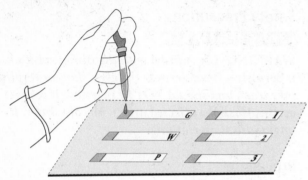

Figure 2

Data and Observations

Table 1

Simulated Urine Sample Data						
Test For	Water	Glucose	Protein	Patient 1	Patient 2	Patient 3
Glucose						
Protein						

Classic **Lab** 37, **How healthy are they?** continued

Analyze and Conclude

1. What do the changes you saw in Part A of the laboratory tell you about the patients and the original solutions? Explain.

_____

_____

_____

_____

_____

2. What do the changes you saw in Part B tell you about the patients and the original solutions?

_____

_____

_____

_____

_____

3. Which of the three patients showed a normal urine sample? How do you know?

_____

_____

_____

_____

4. Should one of the patients be tested further for diabetes? Which one? How do you know?

_____

_____

_____

_____

_____

5. Are any of the patients showing signs of kidney disease? Explain.

_____

_____

_____

6. Error Analysis What are possible sources of error in your experiment?

_____

_____

_____

_____

7. Is the data gathered here enough for a doctor to make a diagnosis? Should the doctor explore further with more tests? Explain your answer.

_____

_____

_____

_____

Inquiry Extensions

1. Do diet and timing have an impact on these test results? Conduct research online or in the library to see how test results are affected when food has been ingested shortly before the tests. Present your findings to the class in the form of a public service brochure.

2. What other tests can be done on urine? Find an example of another common test involving urine and write the procedure for an experiment to test urine for that condition.

Design Your Own
Lab 38

How do you digest protein?

You probably eat protein every day. Protein is found in such foods as eggs, meat, poultry, fish, dairy products, nuts, beans, and lentils. The body uses protein for tissue growth and repair. Proteins are digested in the stomach by digestive chemicals and a mechanical process that occurs as the stomach churns. Pepsin is the enzyme found in digestive juices that chemically digests the proteins in foods by breaking them into shorter chains of amino acids. Pepsin is most effective in the acidic environment of the stomach.

Problem
Design an experiment that determines what conditions are needed for the digestion of proteins in the stomach.

Objectives
• Design an experiment.
• Compare conditions for the function of pepsin in the digestive process.
• Collect and interpret data, and draw conclusions about the conditions within the stomach.

Safety Precautions

WARNING: *Excercise caution when handling hydrochloric acid.*

Possible Materials
test-tube rack
2% pepsin solution
blue litmus paper
boiled egg white or firm tofu
plastic knife
ruler
graduated cylinder
test tubes with stoppers

marking pencil
2% hydrochloric acid solution
stirring rod (glass)

Hypothesis
Use what you know about digestion and proteins to write a hypothesis that explains the conditions that could accelerate digestion in the stomach and why this would occur.

_____

_____

_____

_____

Laboratory Manual

Design Your Own **Lab** 38, **How do you digest protein?** continued

Plan the Experiment

1. Read and complete the lab safety form.
2. Decide on a source of protein to test.
3. Decide on a procedure for testing the effects of acidic conditions on the digestion of protein. Your experiment might take two days to complete.
4. Identify the independent variable, dependent variable, constants, and control group.
5. Decide how you will record your data. Design a data table to record the information that you collect.

Check the Plan

1. Be sure that a control group is included in your experiment and that the experimental groups only vary in one way.
2. Make sure your teacher has approved your experimental plan before you proceed.
3. When you have completed the experiment, dispose of the liquid as directed by your teacher.

Record the Plan

In the space below, write your experimental procedure and make a sketch of your experimental setup.

Data and Observations

1. Use the space below to create a data table of your findings, including the appearance of the protein source over two days.

Analyze and Conclude

1. Which chemicals were best at digesting the protein you examined? How do you know?

_____

_____

_____

2. Does the chemical digestion of protein happen quickly or slowly? Explain.

_____

_____

_____

3. Did you cut the pieces of protein into blocks the same size? Why would that be important?

_____

_____

_____

4. What did your experiment show about the ability of pepsin to digest protein?

_____

_____

_____

5. Error Analysis What were some possible sources of error in your experiment?

_____

_____

_____

6. Describe the control in your experiment. What did the control show?

_____

_____

_____

7. Exchange your procedure and data with another group in the class for peer review. What do their data indicate about the conditions affecting protein digestion?

_____

_____

_____

Write and Discuss

Write a short paragraph describing your findings and indicating whether or not they support your hypothesis. Discuss any questions your results might have raised.

_____

_____

_____

_____

Inquiry Extensions

1. Does digestion occur faster if the particles of protein are smaller? Design a test which examines how the size of particles impacts digestion rate.

2. What other factors could impact the rate of digestion? Would the presence of other food in the stomach play a role? Would more liquid accelerate the process? Choose one additional variable to test, and design an experiment to test it.

Classic Lab 39

How does a body grow?

As the human body grows and develops, it increases in size. The proportions of some parts of the body also change in relation to each other. Males and females develop at different times and at different rates. As you complete this lab, you will discover some proportional relationships of body parts evident in different ages.

Objectives

- Compare the average height of humans with the average length of body parts during various stages of human development.
- Graph the average height of humans.
- Analyze rates of human development, male or female, based on the data assembled and graphed.

Materials

metric ruler
graph paper
calculator
colored pencils

Procedure

Part A. Growth and Development in Humans

1. Read and complete the lab safety form.
2. **Figure 1** shows four stages of development in the life of a human. Note how some parts of the body are larger in proportion to the entire body at various stages of development.
3. **Table 1** shows the average height of a male and female at each of these four stages of human development. Keep in mind that these are the median heights—about half of all people are taller, and about half are shorter.

Figure 1

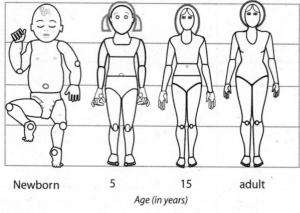

Newborn   5   15   adult
Age (in years)

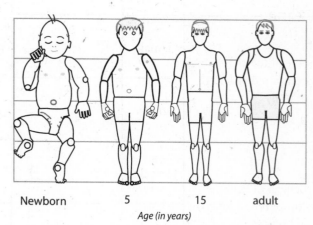

Newborn   5   15   adult
Age (in years)

Differential growth rates in humans. Chart shows head, torso and leg in relation to puberty from birth to adulthood.

Table 1

Average Height at Selected Ages				
	Newborn	**Five Years**	**Fifteen Years**	**Adult**
Average height : male (in cm)	50	109	170	177
Average height: female (in cm)	47	107	162	163

Part B. Graphing Proportions of Body Parts to Total Height

1. Gather information about how the lengths of certain body parts in each of the four stages of development compare to the total length of each body in **Figure 1**. Use your ruler to measure the total length of each figure and the length of the head and torso. (The torso is the part of the body from the shoulders to the pelvis.)

2. Use your calculator to convert your measurements into percentages. Complete **Table 2** with the information you have obtained.

3. Use the graph paper to make a line graph of the data in **Table 2**. Let the *x*-axis be percentage of total body length, and the *y*-axis be age. Use different colored pencils for the head, torso, and leg measurements at each age.

Data and Observations

Table 2

Data for Graphing Proportions of Body Parts to Total Height					
		Newborn	Five Years	Fifteen Years	Adult
Size of head compared to length of body	female				
	male				
Size of torso compared to length of body	female				
	male				
Length of legs compared to length of body	female				
	male				

Classic **Lab 39, How does a body grow?** continued

Analyze and Conclude

1. How do the proportions for head size, torso size, and leg length change from infancy to adulthood?

_____

_____

_____

2. What can you conclude about the rate at which different parts of your body grow? Explain.

_____

_____

_____

3. Was it useful to represent the data from this lab in the form of a table and in a line graph? Explain.

_____

_____

_____

4. Based on your findings, how would you conclude that other measurements of body parts to total body length, for instance, head circumference or arm length, might change as a person grows?

_____

_____

_____

5. Based on your findings, if you were given the head, torso, and leg proportions of an unseen individual whose growth pattern was within normal parameters, do you think you could tell if that person was an infant, a child, a teen, or an adult? Explain.

_____

_____

_____

_____

Classic **Lab 39, How does a body grow?** continued

6. Based on your findings, if you were given the age, head, torso, and leg proportions of an unseen individual whose growth pattern was within normal parameters, do you think you could make a reasonable estimate of that person's height? Explain.

_____

_____

_____

7. **Error Analysis** What are possible sources of error in this exercise?

_____

_____

_____

Inquiry Extensions

1. How much do ratios differ if people are taller or shorter than the median heights shown in **Table 1**? Develop a plan to obtain additional images that you can measure and compare ratios as in **Table 2**.

2. Based on your age, can you predict what your own body ratios will be? Use the calculations here to make an estimate of the ratio of your head size, torso size, and leg length. Then, ask another classmate to measure your height and the corresponding measurements. Determine if your estimates were correct. Report your findings to the class.

Classic
Lab 40

Who needs a banana peel?

Bacteria are present everywhere. Many bacteria can cause disease and decay, but in most cases, you remain disease-free because your skin serves as a barrier between the bacteria and the more delicate tissues inside. The peel on a banana is similar to the skin on your body. In this lab, you will test the effectiveness of a banana peel in preventing the decay of the fruit.

Objectives

- Prepare and observe bananas over a period of five days.
- Model the skin's defense against disease using banana peels.
- Conduct an experiment controlling variables.
- Form a conclusion about the necessity of washing and cleaning cuts to prevent disease.

Materials

sealable plastic bags (4)
fresh bananas (4)
rotten banana
permanent marker
water
paper towel
toothpick
cotton swab
rubbing alcohol

Safety Precautions

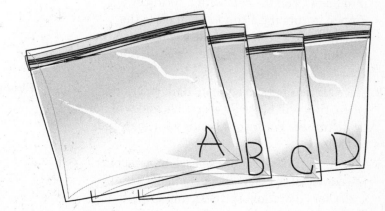

Procedure

1. Read and complete the lab safety form.
2. Label the four plastic bags with your name and the letters A, B, C, and D.
3. Wash the fresh bananas, and then dry them with a paper towel.

4. **Figure 1** Place one banana in Bag A. Seal the bag, and set it aside.

5. **Figure 2** Gently insert a toothpick through the peel of the rotten banana and into the fruit. Then, using the same toothpick, lightly run the tip of the toothpick down the side of the second fresh banana. Do not pierce the peel of the second banana, as shown in **Figure 3**. Use caution when handling the toothpick; it could pierce your skin. Repeat three times, on different parts of the banana. Place the second banana in Bag B, and seal the bag. Discard the toothpick.

6. Take a new toothpick. Insert it into the rotten banana. Then, using the same toothpick, make a shallow, 2.5-cm cut in the peel of the third fresh banana, being careful not to insert the toothpick into the fruit itself. Repeat this three more times, piercing the skin of the banana each time. Place the banana into Bag C, and seal the bag. Discard the toothpick.

7. Use the last fresh banana, and repeat step 6 with one change. Before placing the banana into Bag D, rub the cuts with a cotton swab dipped in rubbing alcohol. Then place the banana in Bag D, and seal the bag. Discard the toothpick and the rotten banana as directed by your teacher.

8. Place all four bags in a warm, dark location where they will be easily accessible and will not be disturbed. Wash your hands with soap and water.

9. Record your observations of each banana over the course of five days in a data table. Set up scales for coloration, softness, and growth of fungi. Use **Table 1** provided on the next page.

10. Every day, for a total of five days, remove the bananas in their bags from storage. Observe the bananas without opening the bags. Record your observations, then return the bags to storage.

11. At the end of the activity, dispose of the unopened bags as instructed by your teacher.

Figure 1

Figure 2

Figure 3

Laboratory Manual

Classic **Lab 40, Who needs a banana peel?** continued

Data and Observations
Table 1

Banana Observation Data				
Day	Banana 1 (no contact with rotting fruit)	Banana 2 (contact with rotting fruit; peel intact)	Banana 3 (contact with rotting fruit; peel pierced)	Banana 4 (contact with rotting fruit; peel pierced, treated)
1				
2				
3				
4				
5				

Classic **Lab 40, Who needs a banana peel?** continued

Analyze and Conclude

1. Why were you asked to pierce the peel of the rotten banana and then use that tooth-pick to scratch the fresh bananas? What did the rotten banana represent?

_____

_____

2. At the conclusion of your observations, how did the appearance of the bananas compare? What other properties did you note? Explain your answer.

_____

_____

_____

3. Which banana was your control? How did the appearance (qualities) of the control banana change over the five days?

_____

_____

_____

4. Error Analysis What are some possible sources of error in your experiment?

_____

_____

5. How is the banana peel in this experiment similar to the skin on your body? What was the purpose of the rubbing alcohol?

_____

_____

_____

6. After performing this experiment, do you think it is helpful to wash and clean cuts and scrapes on your body? Explain.

_____

_____

_____

Inquiry Extensions

1. How else might you conduct this experiment? What materials would be needed?

2. Use this banana model to design an experiment to explain how hand washing can prevent the spread of the common cold virus.